Modelling of the Microfiltration Process

Zhan Wang
Editor

Modelling of the Microfiltration Process

Springer

Editor
Zhan Wang
Department of Chemistry and Chemical
Engineering
Beijing University of Technology
Beijing, China

ISBN 978-981-99-0470-9 ISBN 978-981-99-0471-6 (eBook)
https://doi.org/10.1007/978-981-99-0471-6

This Springer imprint is published by the registered company Springer Nature Singapore Pte Ltd.
The registered company address is: 152 Beach Road, #21-01/04 Gateway East, Singapore 189721, Singapore

Preface

The multiple geometric increases in population, high-intensity industrialized production, and rapid environmental deterioration lead to the increasing shortage of water resources. Today, urban sewage recycling, industrial wastewater reuse, and seawater utilization became the fundamental methods to solve the shortage of water resources. In these methods, it is inevitable to use membrane technology to improve water quality. For example, a microfiltration membrane was used to absorb and intercept bacteria, colloids, and other substances in domestic sewage treatment. Ultrafiltration membrane was employed to realize the separation of macromolecular substances in many domestic sewage treatments. In addition, membrane technology has also achieved good results in treating oily wastewater, heavy metal wastewater, ammonia and amine wastewater, and food industry wastewater. Therefore, membrane separation technology has become a critical high-tech and sustainable development technology to solve the problems of contemporary energy, resources, and environmental pollution because of its low energy consumption, high efficiency, simple process, and low investment. Furthermore, membrane technology plays a significant role in sewage treatment and is becoming "water treatment technology in the twenty-first century."

However, the fine microfiltration or colloidal substances in the feed will enter the membrane pore, adsorb in the inner surface of the membrane, block the membrane pore, and form a colloidal layer and cake layer on the surface of the membrane in the membrane separation process. Therefore, membrane fouling will deteriorate separation performance and reduce service life. Although the pretreatment of feed solution/suspension, the optimization of process conditions, and the optimal design of membrane modules can significantly improve the membrane fouling, the membrane fouling will come sooner or later. Therefore, how to carry out a membrane cleaning scientifically is of great practical significance. The establishment of mathematical models after the in-depth research on the membrane fouling/cleaning mechanism in the filtration/cleaning process will lay a theoretical foundation for the scientific development of membrane cleaning operation.

In 1991, as a student of the China Scholarship Council, I studied for a doctorate at the Saint Petersburg Technological Institute (Technical University) in Russia. While studying in Russia, I first came into contact with the membrane and was attracted by its magical functions. From then on, I began to learn membrane technology and establish a membrane process model. Later, after returning to the Beijing University of Technology, I continued my research and built some models of flat membrane devices. Then, as a visiting research professor from March 2009 to March 2010, I went to Professor George Belfort's team at the Rensselaer Institute of Technology in the United States to further study membrane technology. These accumulated experiences in Russia and the United States provide the basis for a subsequent study on the issue of membrane fouling. Today, our team has established more than 100 mathematical models of various types, including 50 theoretical models (21 models for the constant pressure filtration process, 8 models for the constant pressure cleaning process, 21 models for constant flow filtration and cleaning process, 7 empirical models) and 50 machine learning models (45 regression models, 3 Neural Network models, and 2 Support Vector Machine models). Furthermore, our team has also become one of the few research groups in the world specializing in the model building in the membrane filtration and cleaning processes. In addition, my monograph, "Fundations of Membrane Separation Technology," has been reprinted three times. It was cited by 300 professional books and more than 2000 articles and collected by nearly 900 libraries. However, we found that researchers need to know the pulse of model development when reading many models, which will consume a lot of time and energy for each researcher. Therefore, editing a professional book on the membrane model is urgent. The Chap. 1 is basic knowledge, edited by Professor Wang Zhan of Beijing University of Technology; Chap. 2 is the fouling mechanism and corresponding filtration model of microfiltration and ultrafiltration commonly used under constant pressure/flow filtration mode, which is edited by a Ph.D. student Hu Ge and Prof. Wang Zhan at Beijing University of Technology. Chapter 3 is the mathematical model of the cleaning process (water cleaning, backwashing, chemical cleaning, and ultrasonic cleaning), edited by master's degree student Gao Shujuan and Prof. Wang Zhan at Beijing University of Technology. Chapters 4–7 edited by Prof. Wang Zhan at Beijing University of Technology. We also take the opportunity to pay our efforts to Wang Shumei, Zhang Xinmiao, Liu Dezhong, Zhou Chong, Liu Mei, Wu Wenjuan, Chu Jinshu, Yao Jinmiao, Cui Yanjie, Zhou Yuenan, Yang Liying, Zhao Shanshan, Song Ying, Xi Xuejie, Dong Wenyue, Zhu Zhongya, Gao Kui, Zhang Ximing, Kong Yadong, Li Yanling, Hou Lei, Yao Wei, Du Xinpei, Gao Zhenfu, Guo Yangyang, Li Qiang, Luo Huijia, Gao Shujuan, Qiao Zhaoyu, Hu Ge, Zhong Chenyin, Liu Xinsen, Li Xuwei, Guo Jinyan, Liu Huihui, Lin Xi, Zhou Xingyuan, Wang Guanzhong, Lian Yueru, Natsagdorj Khaliunaa, Khurram Rooha and other students expressed their gratitude for their contributions to the research group. We are very grateful to Springer Verlag Publishing Group and the editors for their hard work and dedication in publishing this book. Without their earnest efforts, this book cannot be published. Although this book cannot cover all the existing mathematical models in the microfiltration process, it takes the main models and the primary historical process and pulse of model development as the main line to help readers get familiar with the development of the model in the microfiltration process as soon as possible. In the editing process, many models are referenced, which also deepens our understanding of the process and models. Although we could not thank every author, we would like to express our sincere thanks to all colleagues in the field of membrane separation and thank them for supporting our research team in the past.

Beijing, China Prof. Zhan Wang

Contents

Basic Knowledge

1 Accumulated Permeate Volume (m^3)

The volume of the permeate (m^3) collected within a specific filtration time (t).

2 Membrane Flux ($m^3/(m^2 \cdot s)$)

$$J = \frac{\Delta V}{A \cdot \Delta t} \tag{1}$$

where J is the membrane flux ($m^3/(m^2\ s)$); ΔV is the volume of the permeate (m^3) collected within the filtration interval Δt and A is the effective filtration area of the membrane.

3 Resistance (m^{-1})

The total membrane resistance can be divided into the virgin membrane resistance, the concentration polarization resistance, the cake resistance, and the pope resistance.

3.1 The Concentration Polarization Resistance (m^{-1})

In the membrane filtration process, solutes and large particles are rejected, while solvents and water pass through the membrane. This way, the concentration of rejected solutes or large particles on the membrane (C_s) will be greater than the concentration of solutes or large particles in the bulk stream (C_b). This phenomenon is called the concentration polarization phenomenon. The resistance caused by

Z. Wang (ed.), *Modelling of the Microfiltration Process*,
https://doi.org/10.1007/978-981-99-0471-6_1

the concentration polarization phenomenon is called concentration polarization resistance.

3.2 *The Virgin Membrane Resistance* (m^{-1})

J_0 is the DI-water flux of the virgin membrane ($m^3/(m^2 \cdot s)$), which can be used to calculate the virgin membrane resistance (R_m):

$$J_0 = \frac{\Delta V}{A \cdot \Delta t} = \frac{\Delta P}{\mu R_m} \tag{2}$$

where ΔV is the volume of the permeate (m^3) collected within filtration interval Δt; A is the effective filtration area of the membrane; ΔP is trans-membrane pressure (TMP) (Pa), and μ is the viscosity of the permeate (Pa·s).

3.3 *The Cake Resistance* ($R_c(t)$) (m^{-1})

In the membrane filtration process, solutes, large particles, and colloidal substances will be rejected by the membrane, and a cake will be formed on the surface of the membrane. The resistance of the cake formed on the membrane surface is called cake resistance.

3.4 *The Pope Resistance* ($R_p(t)$) (m^{-1})

The resistance is related to the shrinkage and blockage of the membrane pores.

4 The Specific Resistance

Generally speaking, the specific resistance can be divided into the specific resistance per unit thickness and the specific resistance per unit mass.

4.1 *The Specific Resistance Per Unit Thickness (m^{-2})*

$$R_c = r \cdot L \tag{3}$$

where r is the specific resistance per unit thickness (m^{-2}), and L is the cake thickness.

4.2 *The Specific Resistance Per Unit Mass (m/kg)*

$$R_c = \alpha \cdot M \tag{4}$$

where α is the specific resistance per unit mass (m/kg), and m is cake mass per unit area (kg/m^2).

5 The Flux Recovery Ratio (*FRR*)

In constant pressure mode, the flux recovery ratio (*FRR*) for the fouled membrane at the hydraulic cleaning time of t can be determined:

$$FRR = \frac{J(t)}{J(0)} \times 100\% \tag{5}$$

where J(t) ($m^3/(m^2$ s)) is the pure water flux of the fouled membrane at hydraulic cleaning time of t and *J(0)* is the pure water flux of the virgin membrane ($m^3/(m^2$ s)).

6 The Trans-Membrane Pressure Recovery (*PRR*)

In constant flow rate mode, the trans-membrane pressure recovery (η) is defined as [1, 2]

$$\eta = [(P_b(0) - P_b(t))/(P_b(0) - P_m)] \times 100\% \tag{6}$$

where P_b (0) (kPa) is the trans-membrane pressure at the backwash time of 0; $P_b(t)$ (kPa) is the trans-membrane pressure at the backwash time of t and P_m (kPa) is the trans-membrane pressure for the virgin membrane before the filtration process, respectively.

7 The Hydraulic Cleaning Efficiency (*CE*)

$$CE = \left(1 - \frac{R_f(t)}{R_f(0)}\right) \times 100\% \tag{7}$$

where $R_f(t)$ and R_f (0) are the fouling resistance (m^{-1}) at hydraulic cleaning time of t and 0, respectively.

8 The Filtration Efficiency (*P*) ($m^3 \cdot m^{-2} \cdot s^{-1}$)

It is defined as the net volume of the permeate during a unit time and area [3–6]:

$$p = \frac{\sum V_f - \sum V_w}{(\sum t_f + \sum t_w)A_m} \tag{8}$$

where $\sum V_f$ is the accumulated permeate volume (m^3); $\sum V_w$ is the accumulated solution volume consumed in the washing process (rinsing, backwashing, and so on) (m^3); $\sum t_f$ is the accumulated filtration time (s) and $\sum t_w$ is the accumulated washing duration (s).

9 Operation Mode

Two categories for distinguishing membrane processes are widely used. One is to focus on the constant parameter caused by the driving force, usually including constant pressure mode and constant flow rate mode. Another category is according to the feed state, dead-end filtration, and cross-flow filtration.

9.1 Constant Pressure Mode

In this operation mode, the trans-membrane pressure drop across the membrane is constant.

9.2 Constant Flow (Rate) Mode

In this operation mode, the flow rate through the membrane is constant.

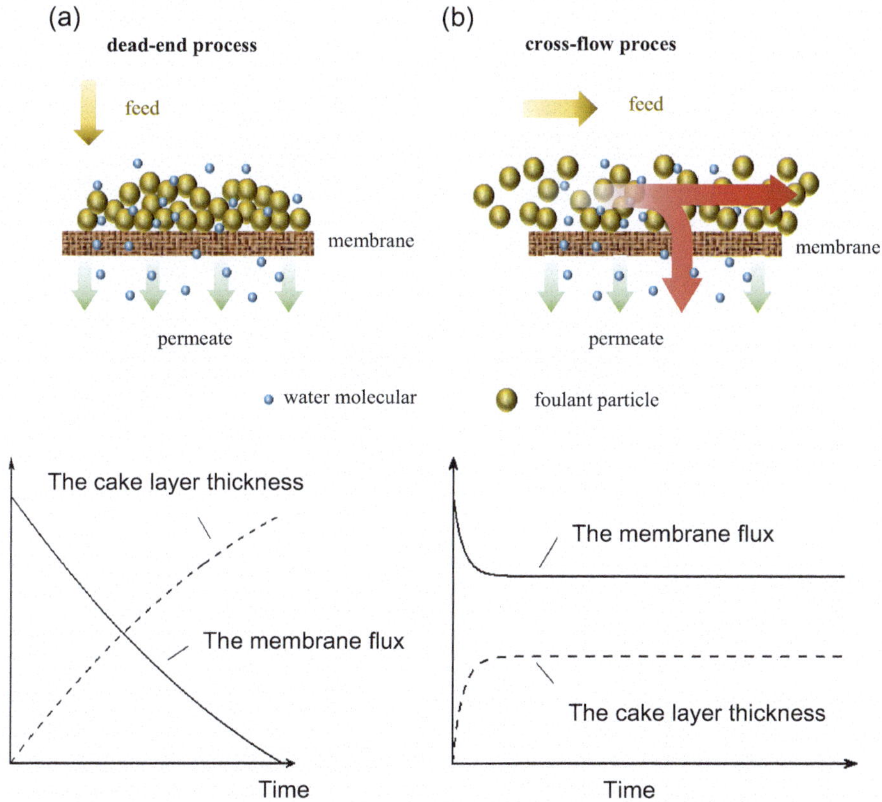

Fig. 1 **a** dead-end process and the corresponding cake layer thickness and membrane flux **b** cross-flow process and the corresponding cake layer thickness and membrane flux

9.3 Dead-End Filtration

The flow direction is perpendicular to the membrane surface and the permeate is through the membrane by the trans-membrane pressure drop. The schematic diagram is shown in Fig. 1a.

9.4 Cross-Flow Filtration

The feed is divided into two parts. One part is perpendicular to the membrane surface and the permeate is through the membrane by the trans-membrane pressure drop while another part is parallel to the membrane surface. This tangential flow with velocity will introduce shear stress to remove the depositing and already deposited

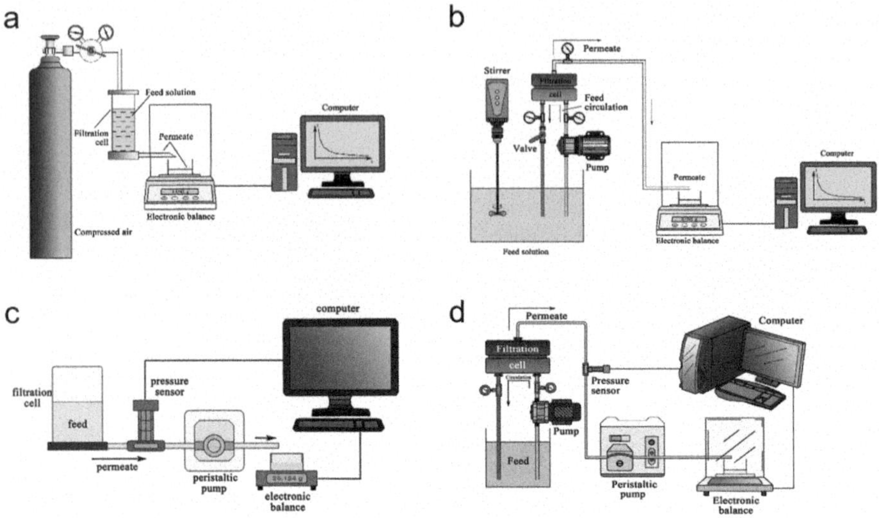

Fig. 2 Schematic diagram of membrane filtration device in lab scale of constant pressure dead-end filtration mode [7] (**a**), constant pressure cross-flow filtration mode [7] (**b**), constant flow dead-end filtration mode [8] (**c**), constant flow cross-flow filtration mode [9] (**d**)

foulants, thus avoiding excessive thicknesses of the cake layers. The schematic diagram is shown in Fig. 1b.

By choosing whether the constant parameter is pressure or flow rate, and whether the feed flow direction is vertical to the permeate direction or not, the membrane filtration can usually be divided into four cases: (1) constant pressure dead-end filtration mode; (2) constant pressure cross-flow filtration mode; (3) constant flow rate dead-end filtration mode, and (4) constant flow rate cross-flow filtration mode, whose schematic diagram of a device in lab scale are shown in Fig. 2.

10 Shear Stress

For Newtonian fluid, the shear stress is proportional to the shear rate:

$$\tau = \mu \frac{du}{dy} \tag{9}$$

where τ is the shear stress(N/m^2) and μ is the viscosity of the fluid (Pa $\cdot$ s).

In stirred cup filter, the average shear stress on the membrane surface can be calculated by the average value of inner and outer regions, and it can express as follows [10]:

$$\overline{\tau} = \frac{6.6\frac{\mu\omega}{\delta}r_c^{1.6}\left(\frac{1}{3}r_c^{1.4} + \frac{5}{7}\left(\left(\frac{D_t}{2}\right)^{1.4} - r_c^{1.4}\right)\right)}{D_t^2} \tag{10}$$

where μ is viscosity (Pa · s); ω is angular velocity (rad/s); r_c is the critical radius (m); δ is the momentum boundary layer ($\delta = \sqrt{\mu/\rho\omega}$), and D_t is the filtration cell diameter (m).

11 Critical Flux

The critical flux reflects an ideal filtration state without the membrane fouling in the operation process. When the operation flux below the critical flux, neither particles are adsorbed to the inner wall of the membrane pore and the membrane surface, nor the formation of concentration polarization layer. The operating flux has a linear relationship with the trans-membrane pressure. In contrast, if the operation flux is higher than the critical flux, there will be the adsorption and deposition of particles on the membrane surface and in the membrane pores, and then form a concentration polarization layer and cake layer. In this case, the relationship between the operation flux and trans-membrane pressure is nonlinear and deviates from the straight line.

12 Diffuse Coefficient

The diffusion coefficient at different temperatures can be estimated by the following equation:

$$D = 2.5 \times 10^6 \frac{k_B T}{\mu(M_W \cdot [\mu])^{1/3}} \tag{11}$$

where k_B is the Boltzmann constant (1.38×10^{-23} J/K); T is the temperature (K); μ is the dynamic viscosity of fluid (Pa·s); $[\mu]$ is the intrinsic polymer viscosity, and M_W is the molecular weight of the solute.

13 Three Basic Approaches for Establishing Model

13.1 Empirical Method

The empirical formulas were used to describe the variation of some parameters in the membrane process.

13.2 Machine Learning Method (Black-Box Model)

Machine learning (ML) is an interdisciplinary subject, involving probability theory, statistics, convex analysis, algorithm complexity theory, and other disciplines. In the membrane field, machine learning is mainly used to establish mathematical model by using regression method, neural network, and support vector machine.

13.3 Mathematical Method

Based on the deep understanding of the process mechanism, the complex actual process is reasonably simplified, the physical model for the specific problem to be solved is abstracted, and then it is mathematically described. In this process, mass conservation and momentum conservation are mostly needed and the model parameters need to be determined by later experiments.

References

1. J.C.T. Lin, D.J. Lee, C.P. Huang, Membrane fouling mitigation: membrane cleaning **45**, 858-872 (2010)
2. H.Q. Chang, H. Liang, F.S. Qu, B.C. Liu, H.R. Yu, X. Du, G.B. Li, S.A. Snyder, Hydraulic backwashing for low-pressure membranes in drinking water treatment: a review. J. Membr. Sci. **540**, 362–380 (2017)
3. K.J. Hwang, C.S. Chan, K.L. Tung, Effect of backwash on the performance of submerged membrane filtration. J. Membr. Sci. **330**, 349–356 (2009)
4. L. Yang, Z. Wang, Y. Sun, Z. Hu, S. Zhao, X. Wang, W. Li, X. Xi, J. Zhang, Z. Pei, Influence of various operating conditions on cleaning efficiency in sequencing batch reactor (SBR) activated sludge process. Part II: Backwash and water rinsing introduced membrane filtration process. Desalination **272**, 76–84 (2011)
5. M. Kennedy, S.M. Kim, I. Mutenyo, L. Broens, J. Schippers, Intermittent crossflushing of hollow fiber ultrafiltration systems. Water Supply **118**, 175–187 (1998)
6. Y.H. Gao, J. Qin, Z.W. Wang, S.W. Osterhus, Backpulsing technology applied in MF and UF processes for membrane fouling mitigation: a review. J. Membr. Sci. **587**, 20 (2019)
7. G. Hu, C.Y. Zhong, Z. Wang, X. Wang, Iop, Application of error function in dead-end/cross-flow microfiltration Mode, in *6th International Conference on Advances in Energy Resources and Environment Engineering (ICAESEE)* (Iop Publishing Ltd, Chongqing, Peoples R China, 2020)
8. S. Yin, P. Song, Y.J. Chen, X.P. Du, X.G. Liu, H. Zhang, Z. Wang, X. Wang, Iop, An investigation of the membrane fouling behaviors in constant flux mode, in *6th International Conference on Advances in Energy Resources and Environment Engineering (ICAESEE)* (Iop Publishing Ltd, Chongqing, Peoples R China, 2020)
9. G. Hu, X. Liu, Z. Wang, X. Du, X. Wang, Comparison of fouling behaviors between activated sludge suspension in MBR and EPS model solutions: a new combined model. J. Membr. Sci. **621** (2021)
10. N. Shamsuddin, D.B. Das, V.M. Starov, Filtration of natural organic matter using ultrafiltration membranes for drinking water purposes: circular cross-flow compared with stirred dead end flow. Chem. Eng. J. **276**, 331–339 (2015)

Constant Pressure Mode

Nomenclature

$A(t)$	The available membrane area at filtration time t
a'	Regression coefficient in Eq. (6)
A'	Coefficient in Sect. 2.2.7
A''	Coefficient in Sect. 2.2.8
A_0	The initial membrane area
$a_0(x)$	The dimensionless pressure component under steady state
$a_1(x)$	The dimensionless pressure component under unsteady state
A_{agg}	A given membrane area that an aggregate can block
$A_{blocked}$	The blocked membrane area at filtration time t
A_m	Relative amplitude of fluid
A_{open}	The membrane area at filtration time t
B	Parameter in Sect. 2.2.7
b	Parameter in Eq. (3)
b'	Regression coefficient in Eq. (6)
b''	Constant in Sect. 3.1
c	The volume of particle trapped per unit filtrate volume per unit membrane area (Eq. (16))
C	Concentration of solution/suspension
C'	Concentration of feed at the boundary layer in Eq. (8)
c'	Regression coefficient in Eq. (6)
C''	Parameter in Sect. 2.2.4 ($C'' = (1 + \Delta p_c/p_a)1 - \delta - 1)$
C_0	Initial bacteria concentration
C_b	Concentration of bulk solution
C_f	Effective drag coefficient
C_m	Concentration at the membrane surface
C_p	Concentration in permeate
CT	Contribution of T to S
D	Diffusion coefficient

Z. Wang (ed.), *Modelling of the Microfiltration Process*,
https://doi.org/10.1007/978-981-99-0471-6_2

D'	Constant in Eq. (6)
d_p	Fouling particle diameter
d_r	The variation of pore radius in a tiny time dt
ET	Dependent variable in Eq. (6)
F	Faraday constant (96485 C mol^{-1})
f	The fractional amount of total protein present as aggregates
f'	The fractional amount of protein contributed to deposit resistance
F_{AB}	Lewis acid-base force
F_{EL}	Electrostatic double-layer force
F_g	Net gravitational force
F_I	Total interaction force
F_{LW}	Lifshitz-van der Waals force
f_n	Pore probability density functions
F_p	Drag force induced by permeate flow
f_s	Solute probability density functions
g_c	Conversion factors
h	Separation distance in Sect. 2.2.9
h_0	The minimum separation distance (0.158×10^{-9} m)
h_f	Half height of membrane channel in Sect. 2.2.8
h_m	The channel height of filtration module
J	The permeate flux
J_0	The initial permeate flux
$J(x)$	The filtration flux along the x-axis
$J_{blocked}$	The permeate flux through blocked pores
J_{c0}	The initial filtration flux in colloid cake formation stage
k	The cake permeability in Sect. 2.2.4
K	A corrected parameter in Sect. 3.4
K'	Coefficient in Sect. 2.2.7
K''	Coefficient in Sect. 2.2.8
K''	Parameter in Eq. (4)
k'	Deposition factor
k_0	Cake permeability at the zero-stress state
k_1	Constant in Eq. (52)
k_2	Denudation coefficient
K_b	Complete pore-blocking parameter in Sect. 3.3
k_c	The proportional constant
K_c	The cake filtration constant ($\mu \alpha_{av} \rho_s/(p(1-ms))$)
K_{cc}	Colloid cake layer filtration constant
K_d	Cake equilibrium coefficient
K_{fc}	Flocs cake filtration constant
K_h	Hole constriction constant
K_i	The intermediate pore-blocking constant
K_s	The blocking constant for standard pore-blocking mechanism $\left(2c/\left(\pi L N_0 \left(1-\varepsilon_p\right) r_0^2\right)\right)$

$\overline{k}$	Corrected parameter in Eq. (50)
L	The length of the membrane pore
l	Total cake thickness
$L^{'}$	The length of the channel of filtration module
m	The ratio of the mass of wet to the mass of dry cake
M	The cake mass
m/A	bacteria mass in the cake per unit membrane area
M_{agg}	Mass of a single aggregate
m_p	Mass of protein deposit onto the membrane surface
n	The cake compressibility in Eq. (59)
$n^{'}$	Parameter in Eq. (2)
$n^{''}$	Experiential constant in Sect. 2.2.7
$n(r_p,t)$	Pore concentration density
N_0	The number of open pores per unit effective membrane area for a virgin membrane
$N_0^{'}$	The number of open pores per unit effective membrane area at filtration time t
$N_{blocked}$	The quantity of blocked pore $\left(= N_0 - N_0^{'}\right)$
ΔP	The trans-membrane pressure drop
p	The trans-membrane pressure drop
$P(x)$	The pressure along the fiber at distance x from the entrance of the hollow-fiber module
P_0	Atmospheric pressure
p_a	The normalizing parameter of compressive pressure
p_l	The pressure the liquid in cake undertaken
p_s	The solid compressive pressure
Q	The feed volume flow in Sect. 2.2.8
$q(x)$	The volume flow rate along the fiber
Q_0	The initial volumetric flow rate
$\overline{Q}_0$	The average feed volume flow in Sect. 2.2.8
q_l	Liquid apparent rate
$q_{l\|L^-}$	Liquid velocity in the cake phase
$q_{l\|L^+}$	Liquid velocity in the suspension phase
Q_{max}	The maximum feed volume flow
Q_{min}	The minimum feed volume flow
Q_{open}	The volumetric flow rate through the open pores
q_s	Solid apparent rate
Q_{ss}	The quasi-steady flux through the completely fouled membrane
r	Membrane pore radius at time t
R	Filtration resistance
$R^{'}$	Specific protein resistance in Sect. 3.2
R_0	The membrane resistance
$R_1{'}$	The separation media resistance
$R_1^{''}$	The resistance of any throttling layer
R_2	The cake resistance

R_3	The pipe resistance of the filtration system
R_4	The resistance of any valves
Ra	The adsorption resistance
R_{ass}	Steady value of adsorption resistance
r_c	The thickness specific cake resistance
R_c	The cake resistance
r_d	The specific cake resistance per unit thickness
$R_d(t)$	The average resistance of foulant deposit on the inside surface of the hollow fiber
$R_d(x)$	The resistance of foulant deposit along the hollow fiber
Re	Reynolds number
R_{in}	The internal radius of the hollow fiber
R_m	The virgin membrane resistance
r_p	Pore radius
R_p	The resistance of proteins deposit over a particular region
$R_p{}'$	The concentration polarization resistance
R_{pss}	The steady concentration polarization resistance
r_s	Solute radius
S	Dependent variable in Eq. (6)
s	The mass fraction of feed
SM	Dependent variable in Eq. (6)
t	Filtration time
T	Dependent variable in Eq. (6)
t_1	Transition time between initial filtration stage and earlier pseudo-steady stage in Sect. 3.5
t_2	Transition time between earlier pseudo-steady stage and late pseudo-steady stage in Sect. 3.5
t_p	The time that the region was covered or blocked by proteins firstly
t'_p	Time required to reach the transition point of the fouling mechanism
U	Liquid velocity
u	The flow velocity along the x-axis in Sect. 2.2.8
$u(t)$	The average flow velocity of the feed solution/suspension in the hollow fiber
$u_0(y)$	The dimensionless velocity component under steady state
$u_1(y)$	The dimensionless velocity component under unsteady state
$\overline{u}$	The average flow velocity in Eq. (62)
v	The flow velocity along the y-axis in Sect. 2.2.8
V_0	Initial retentate volume
V_m	Virtual permeate volume
v_w	The flow velocity near the membrane surface
w	Cake mass
w_c	Virtual mass
x	The distance from the membrane surface
x'	The number of pores blocked per unit filtrate volume in Eq. (10)
z	Ion valence

α	The specific cake resistance
$\alpha^{'}$	The pore-blocking parameter in Sects. 2.2.5 and 3.2
α_0	The specific cake resistance at the zero-stress state
α_{av}	The average specific cake resistance
β	Empirical constant in Sect. 2.2.4
$\beta^{'}$	Coefficient in Eq. (78)
$\beta*$	Corrected parameter
β^0	Parameter in Sect. 2.2.5
$\overline{\beta}$	Linear parameter.
Γ	Dependent variable in Eq. (5)
γ^-	Surface tension parameters
γ^+	Surface tension parameters
γ^{LW}	Surface tension parameters
δ	Empirical constant in Sect. 2.2.4
δ_0	The volume of cake formed on the surface of membrane to get unit volume of permeate
δ_c	The cake thickness
$\delta_d(t)$	The average foulant layer thickness on the inside surface of the hollow fiber varies with time
$\delta_d(x)$	The thickness of deposited substance along the x-axis
δ_p	The boundary layer thickness in Eq. (9)
Δp_c	The pressure drop across the filter cake
Δp_m	The pressure drop across the membrane
ε	The cake porosity
ε_e	Dielectric permittivity of water (7.17×10^{-10} C V^{-1} m^{-1}) in Sect. 2.2.9
$\varepsilon_{l\|L^-}$	Volume fraction of particle in the cake phase.
$\varepsilon_{l\|L^+}$	Volume fraction of particle in the suspension phase.
ε_p	The porosity of foulant inside the membrane pore
ε_s	Volume fraction of solid in cake
ε_{s0}	Volume fraction of solid in cake at the zero-stress state
$\overline{\varepsilon}$	Error terms in Eq. (5).
$\zeta(t)$	The average surface wall shear rate in the hollow fiber
Λ	Parameter ($= \pi \Delta P / 8 \mu L$)
λ_{AB}	Characteristic decay length for AB
λ_{EL}	Characteristic decay length for EL
λ_{LW}	Characteristic decay length for LW
μ_1	Suspension viscosity
ρ	The filtrate density
ρ_s	Mass density of solids comprising the cake
σ_f	Parameter of foulant calculated by zeta potential
σ_m	Parameter of the membrane calculated by zeta potential
τ_w	Shear stress
Φ	Index in Eq. (49)
φ_b	The solid volume fractions in suspension
φ_c	The solid volume fractions in cake

φ_w	The particle volume fraction in the boundary layer
ω	Fluid frequency of pulsatory movement
ΔG_{awb}^{LW}	Free energy for Lifshitz-van der Waals
ΔG_{awb}^{AB}	Free energy for Lewis acid–base force

Before Reading

In this section, some simple mathematical theories and tools should be used frequently like calculus, differential equation, higher order differential equation, and so on. The detailed derivation progresses are not provided in this book due to the reader having enough basis and a certain capability to solve these simple problems about integral and differ.

1 Empirical Model

The empirical model was first used in the fields of life sciences and social sciences. Because there are many factors and complex relations, the internal law is seldom known, so the empirical model is often used to determine the relationship between the factors and the objective function. Later, the empirical model has also been widely used in the field of the membrane. It is broadly divided into three categories as follows:

(a) **Purely empirical model**:

Numerous empirical models have proposed to predict the flux decay. For example, exponential function and power function were often used in these empirical models to describe the variation of the membrane flux.

$$J = J_0 \cdot t^{n'} n' < 0 \tag{1}$$

$$J_0 - J = (J_0 - J) \cdot \left(1 - e^{-t}\right) \tag{2}$$

$$J = J_0 \cdot e^{-b't} \tag{3}$$

Error function is, to some degree, similar with the flux decay trend. Hence, Du was inspired by the heat transfer to propose a model using error function to describe the flux decay (Fig. 1) [1]:

$$\frac{J}{J_0} = \operatorname{erf}\left(\frac{k''}{\sqrt{t}}\right) \tag{4}$$

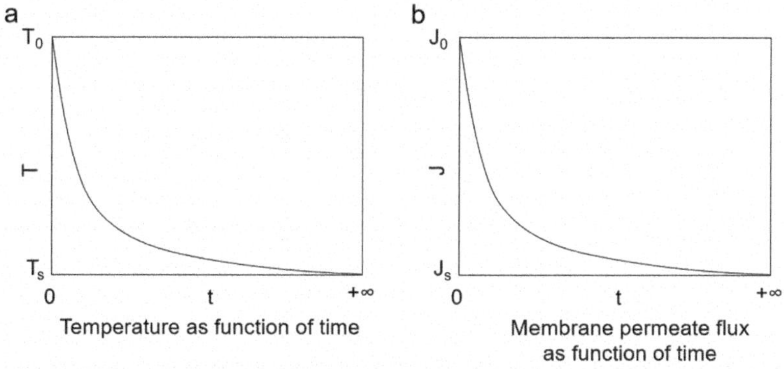

Fig. 1 **a** Application of error function in heat transfer and **b** the variation of the membrane flux with time

(b) Machine learning model (black-box model)

① Regression model

The regression model is widely used in the membrane field to predict the variation of the membrane flux, membrane resistance, the specific cake resistance, etc. as function of filtration conditions or cleaning conditions (Table 1) [2–16]. In essence, regression is a statistical modeling technique used to describe a continuous response variable as a function of one or more predictor variables. According to the function expression describing the causal relationship between independent variables and dependent variables, it can be divided into linear regression and nonlinear regression.

In linear regression, the model describes the relationship between a dependent variable γ (the response) as a function of one or more independent variables Xi (the predictors). The general equation for a linear regression model is

$$\Gamma = \overline{\beta}_0 + \sum (\overline{\beta}_i X_i + \overline{\varepsilon}_i) \tag{5}$$

where $\overline{\beta}$ represents linear parameter estimates to be computed and $\overline{\varepsilon}$ represents the error terms.

In nonlinear regression, some nonlinear regression problem can be transformed into linear regression problem through variable transformation and then treated by linear regression method. According to the principle of least square method, the coefficient value is calculated, and the obtained model is nonlinear regression model. While, some nonlinearizable problem cannot be transformed into linear regression problem. In this case, based on the least square method of regression problem, a mathematical solution of unconstrained extremum problem in optimization method, simplex method, is applied to the extremum problem of minimum sum of squares of error.

In addition, the influence intensity of multiple independent variables on a dependent variable can also be determined in regression method. For example, in order

Table 1 Regression model for the flux, retention, the specific cake resistance, the accumulated filtrate volume, and backwashing efficiency

References	Operation condition	Regression model	The influence intensity of independent variables
[2]	Dead-end, PES-20 flat-sheet UF membrane, polysaccharide of agaricus blazei murill (1–30 mg/L, 0.08–0.22 MPa, 18–46 °C)	Membrane flux: $J = 10.7982 + 54.1310p - 0.5868C + 0.1499T$	The influence degree: concentration (78.3%) > pressure (16.61%) > temperature (5.09%)
		Membrane retention: $R = 0.9551 - 0.2031p + 0.001C - 0.0005T$	The influence degree: pressure (46.86%) > concentration (43.11%) > temperature (10.03%)
[3]	Dead-end, PVDF flat-sheet 0.1 μm membrane. BSA solution (1–5.0 g/L, 0.04–0.16 MPa, 20–30 °C)	Membrane flux: $J = -1.28 + 0.22T + 33.11p - 1.04C$	The influence degree is pressure (69%) > concentration (18%) > temperature (13%)
		Pure water flux of the membrane: $J_w = -6.57 + 0.27T + 133.27p$	The influence of temperature and pressure on the microfiltration membrane flux was different, 2% and 98%, respectively
[4]	Dead-end, PES flat-sheet 0.1 μm membrane, yeast suspension (0.1–3.0 g/L, 0.04–0.14 MPa, 20–50 °C)	Membrane flux: $J = -1.804 \times 10^{-7} + 3.149 \times 10^{-5}p + 3.526 \times 10^{-7}T - 2.573 \times 10^{-6}C + 1.388 \times 10^{-8}\omega$	The influence digress is rotate speed (64.9%) > temperature (14.4%) > pressure (12.9%) > concentration(7.8%), all four factors are remarkable influence factor
[5]	Dead-end, PES flat-sheet 0.2μm membrane, yeast suspension (0.3–2.0 g/L, 0.04 – 0.12 MPa, 20–40 °C)	The specific cake resistance: $\alpha = 1.639883 + 44.2p + 0.86217C - 0.0607T$	The influence digress is pressure (75.31%) > concentration (12.89%) > temperature (11.8%)
[6]	Cross flow, PVD flat-sheet 0.1 μm membrane, yeast suspension (1–5.0g/L, 0.02.0.10 MPa, 20–45 °C)	The specific cake resistance: $\alpha = 1.4704 \times 10^{14} + 4.3652 \times 10^{14}u + 1.2256 \times 10^{15}p + 1.8224 \times 10^{13}C - 2.9467 \times 10^{12}T$	The influence digresses: cross-flow velocity (38.85%) > concentration (28.32%) > pressure (19.34%)> temperature (13.49%), all four factors are remarkable influence factor, the pressure, concentration, and cross-flow velocity play a positive role in the flux, but the temperature plays a negative part

(continued)

Table 1 (continued)

References	Operation condition	Regression model	The influence intensity of independent variables
[7]	Dead-end, PES flat-sheet 0.1μm membrane, 0.04–0.12 MPa, agitation velocity (400-800rpm)	Denudation coefficient (kaolin suspension) $K = -0.43 - 1.16P + 147.23C + 5.13 \times 10^{-4}\omega$	The influencing degree of operating conditions on model parameter K is suspension concentration, agitation velocity, pressure, and the percentages of them are 57.5%, 39.8%, and 2.7%, respectively
		Denudation coefficient (yeast suspension) $K = -51.92 - 271.96p + 1.06 \times 10^{4}C + 7.57 \times 10^{-2}\omega$	The influencing degree of operating conditions on model parameter K is suspension concentration, agitation velocity, pressure, and the percentages of them are 69.8%, 24.7%, and 5.5%, respectively
[8]	Dead-end, PAN 0.1 μm membrane, quality of domestic sewage: (COD 180.6–225.8 (mg /L), NH_3–N 45.9–73.6 (mg /L) TOC 86.5–115.5 (mg /L) pH 7.5–8.0 Turbidity 20–26 (NTU)TMP = 0.03–0.18 MPa dead-end microfiltration	The accumulated filtrate volume $V = 249.64\text{TMP} - 0.056\text{MLSS} + 31.69\text{DO} + 55.06\text{pH} - 5.25T - 8.82\text{HRT} + 16.24$	The percentages of influencing degree of these factors were 34.0%, 22.9%, 15.5%, 13.0%, 7.6%, and 7.0%, respectively
		The COD concentration: $\text{COD} = 0.0073\text{MLSS} - 5.64\text{DO} + 2.86T + 0.099\text{SRT} + 2.48\text{HRT} - 88.82$	The percentages of influencing degree of DO, MLSS concentration, temperature, SRT, and HRT were 5.8%, 63.9%, 13.6% and 14.7%, respectively
		NH_3–N concentration: $NH_3\text{–N} = 0.017\text{MLSS} - 12.99\text{DO} - 28.86\text{pH} + 0.90T + 0.13\text{SRT} + 201.30$	MLSS concentration, temperature, and SRT are directly proportional to the NH_3-N concentration in the filtrate, and their percentages were 40.5%, 7.5%, and 15.7%, respectively. DO concentration and pH showed inverse proportionality, with 9.6% and 26.7%, respectively

(continued)

Table 1 (continued)

References	Operation condition	Regression model	The influence intensity of independent variables
[9]	0.1 μm PVDF, the extracted actual EPS solution was filtered under different TMPs (0.02–0.10 MPa), different temperatures (18–26 °C) and different EPS concentrations (35.14–348.75 mg L^{-1}) in a dead-end MF cell	The cumulative filtrate volume (*CFV*) of actual EPS solution $CFV = -3.7\times 10^{-8}C + 5.79\times 10^{-7}T + 1.27\times 10^{-4}\Delta P$	The sequence of influence degree of operating conditions is the temperature (38.1%) > the TMP (34.8%) > the EPS concentration (27.1%)
[10]	Dead-end MF cell 0.1 μm PES membrane, TOC (31.5–37.8 mg L^{-1}), NH_3–N (9.4–11.22 mg L^{-1}) and COD (34.6–40.7 mg L^{-1})	Backwashing efficiency (*P*) and the backwashing conditions, $P = 6.76TMP + 0.028N + 0.030T + 0.0017t$	The percentage contributions of trans-membrane pressure (*TMP*), backwashing times (*N*), temperature of detergent (*T*), and backwashing duration (*t*) to washing efficiency (*P*) were 39.84%, 7.27%, 50.49%, and 2.39%, respectively
[11]	Dead-end microfiltration cell 0.1 μm PES membrane. The main parameters of the feed suspension into the membrane cell used in the experiments were total oxygen content (TOC) (33.5–38.4 mg L^{-1}), NH_3–N (9.7–10.75 mg L^{-1}), and chemical oxygen demand (COD) (48.2–52.6 mg L^{-1})	First cycle: $CMPFV_1 = -1187.24r + 8.07T + 0.65V + 0.86N + 33{\cdot}93\omega$ Second cycle: $CMPFV_2 = -1774.71r + 12.34T + 0.94V + 1.26N + 50.90\omega$ Third cycle: $CMPFV_3 = -2375.42r + 16.77T + 1.22V + 1.65N + 68.40\omega$ Fourth cycle $CMPFV_4 = -2968.36r + 21.03T + 1.50V + 2.03N + 84.35\omega$	The average contribution of water rinsing conditions on CMPFV was relative flux (38.3%)>detergent temperature (21.6%)>washing times (19.3%) > agitation speed (11.2%)>detergent volume (9.6%)
[12]	0. 1 μm hollow-fiber membrane	The membrane flux: $J = 1.898\times 10^{-6}/(1 - e^{-41469 \times 10-5t -011101})$	Yeast suspension

(continued)

Table 1 (continued)

References	Operation condition	Regression model	The influence intensity of independent variables
		$J = 3.137\times10^{-7}/(1 - e^{-11044\times10-5t-0103760})$	Kaolin suspension
[13]	0.1μm PVDF and 0.2 μm PTFE flat-sheet membrane (0.10–0.16 MPa, 18–30 °C)	The membrane flux: $J_E/J_T = \ln(-0.0045T^2 + 0.2309T)$ $\ln(J/\Delta p) = -3.5767\Delta p + 4.6277$	BSA solution
[14]	Using polysulfone as the base membrane material, a flat ultrafiltration membrane was prepared in orthogonal experiment, and then coated with modified polydimethylsiloxane on its surface to prepare polysulfone silicone rubber composite oxygen-rich membrane	Oxygen enrichment concentration: $C = 28.863 + 0.051\chi_1 - 0.106\chi_2 - 0.11\chi_3 + 0.05\chi_1\chi_2 - 0.05\chi_1\chi_3 - 0.1\chi_2\chi_3 - 0.16\chi_1^2 - 0.193\chi_2^2 - 0.227\chi_3^2$ Permeability coefficient: $P = 6.972 + 0.068\chi_1 + 0.149\chi_2 + 0.008\chi_3 + 0.075\chi_1\chi_2 - 0.008\chi_1\chi_3 + 0.015\chi_2\chi_3 - 0.067\chi_1^2 - 0.073\chi^2\chi_2 + 0.041\chi^2\chi_3$	Air
[15]	0.1 μm PAN and polyvinylidene (PVDF) microfiltration membranes	The resistance due to in-pore fouling: $R_{in}(0) = 2.886\times10^8T + 1.213\times10^{12}p - 2.037\times10^9$ The resistance due to the deposit on the membrane surface: $R_s(0) = -1.141\times10^{12}p + 2.983\times10^{11}$	Sodium alginate solution

(continued)

Table 1 (continued)

References	Operation condition	Regression model	The influence intensity of independent variables
[16]	PVDF/CA microfiltration membrane	Pure water flux: $y_1 = 548178 - 9108x_1 - 28195x_2 - 12177x_6 + 2157x_8$ Average bubble point pressure minus maximum bubble point pressure: $y_2 = -01048 + 01008x_2$ Pure water flux plus average bubble point pressure minus maximum bubble point pressure: $y_3 = 624134 - 28194x_2$	Pure water

to quantify the contributions of independent variables (T, SM, and ET) to dependent variable (S), a linear regression equation to describe the relationship between a dependent variable S (the response) as a function of independent variables (T, SM, and ET) was obtained [17] as follows:

$$S = a' \times T + b' \times \mathrm{SM} + c' \times \mathrm{ET} + D' \tag{6}$$

where a', b', and c' are the regression coefficients; D' is a constant; and S is dependent variable.

Then, take T as example, the contribution of T to S was obtained as follows:

$$\mathrm{CT} = \left(\frac{a \times \Delta T}{\Delta S}\right) \times 100\% \tag{7}$$

where ΔS and ΔT are the tiny changes of S and T, respectively.

The contributions of SM and ET to S could be calculated in the same manner.

② **Artificial Neural Network (ANN) model**

Artificial neural network (ANN) is a new soft computing method suitable for solving the mapping problem between targets from one set to another. Generally speaking, it has the advantage of arbitrarily approaching any highly nonlinear function without predetermining the mathematical model of samples. In short, it only needs to provide a certain learning sample data of a given network, and the system can realize nonlinear approximation and prediction of the objective function through the adjustment of network weights and thresholds (Fig. 2). The prediction ability of artificial neural network generally changes with the number of hidden layers, the number of hidden layer neurons, the transfer function, learning function and learning rate of hidden layer and output layer.

In order to optimize the operating conditions in dead-end microfiltration, a predicting model relating the specific cake resistance to the operating conditions based on the BP neural network, which is to realize a nonlinear mapping from input to target output by continuously adjusting the connection weights and thresholds of each layer, was firstly developed [18]. The best structure of BP neural network model is 3-7-1. The number of hidden layer neurons is 7 (Fig. 3), the learning rate is 0.05 (Table 2), the learning function is traingdx (Table 3), and the transfer function is logsig (Table 4). This optimized network structure is used to predict the specific cake resistance of the test samples and its average relative absolute error was 3.55%. In addition, the structure-optimized BP-ANN model also can be applied to simulate the permeate flux [19], the chemical oxygen demand (COD) concentration of effluent from SBR very well.

③ **Support Vector Machine (SVM) model**

Support Vector Machine (SVM) model only considers input and output, and has nothing to do with the process mechanism. Support Vector Machine (SVM) as

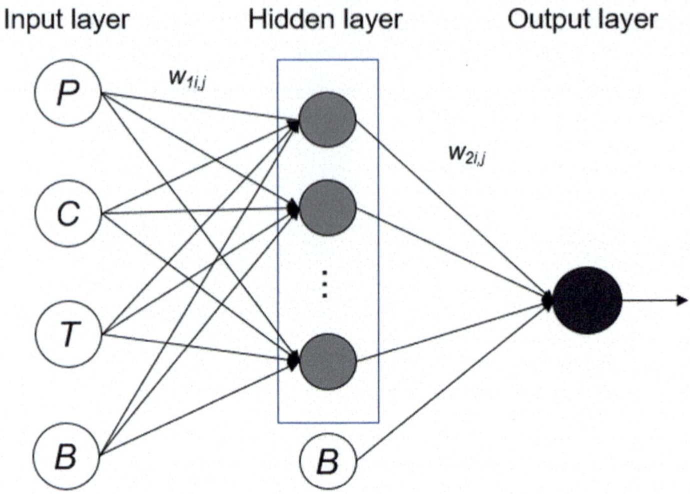

Fig. 2 Sketch of simple structure for ANN. W_{1ij}-weight from input layer to hidden layer; W_{2ij}-weight from hidden layer to output layer and B-bias node

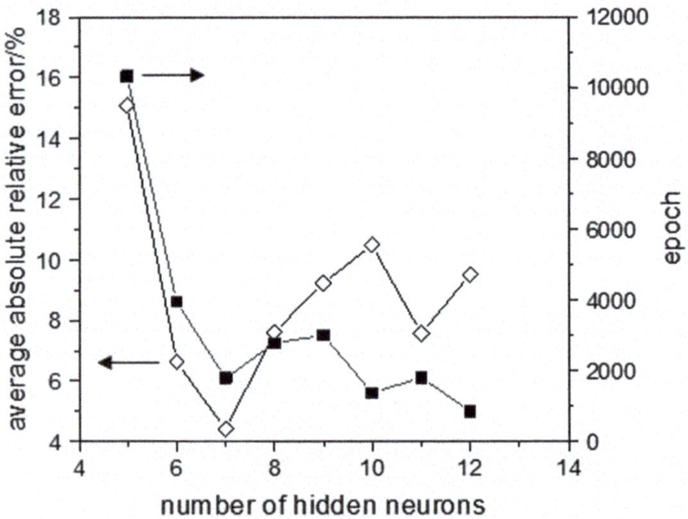

Fig. 3 Average absolute relative error/epoch versus number of hidden neurons

Table 2 Comparison of different learning rates

Learning rate	Epochs	Average relative absolute error/%
0.01	7291	7.12
0.03	7258	7.15
0.05	1764	4.41
0.07	7242	7.38
0.1	7231	7.34

Table 3 Comparison of different training functions

Item	Train function			
	Traingda	Traingdx	Trainrp	Trainlm
Epoch	1764	722	1954	21
Average relative absolute error/%	4.41	3.55	4.99	4.82

Table 4 Comparison of different transport functions

Item	Train function	
	Logsig	Purelin
Epoch	722	8914
Average relative absolute error/%	3.55	4.46

a highly effective approach of system modeling with restricted datasets has been widely applied in many fields for predicting [20]. The detailed relationships between fabrication conditions and filtration performance of membrane were established by the SVM model [21] and the result showed that SVM model shows better learning and generalization ability in the small size of the training set than ANN model. The similar result is also observed for predicting the membrane flux during dead-end microfiltration of activated sludge suspensions from sequencing batch reactor (SBR) with different experimental samples [22] (Fig. 4).

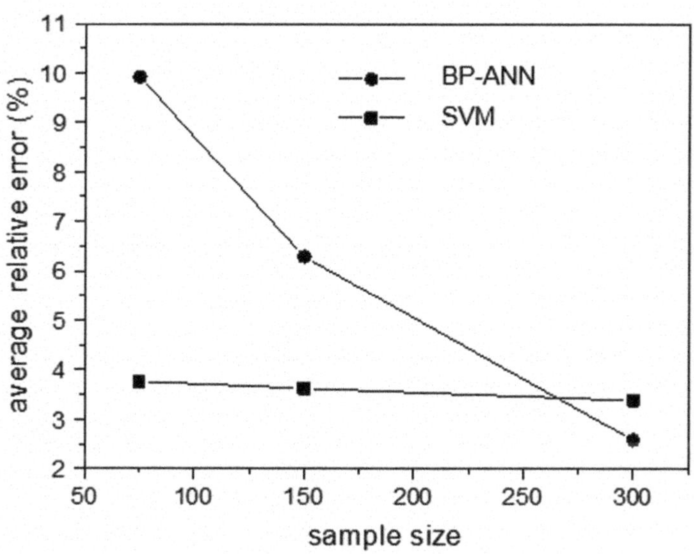

Fig. 4 The average relative errors of the membrane permeate flux of bioreactor by SVM models and BP-ANN model in the case of different sample sizes

2 Theoretical Model

2.1 Concentration Polarization Model

Concentration polarization is a common phenomenon which can also result in flux decay or trans-membrane pressure increase. In fact, concentration polarization is due to the selective permeability of the membrane, solute is rejected at the high-pressure side of the membrane to form a layer with high concentration of solute near the membrane surface, which will induce a back-transport of solute from this layer to the bulk feed solution. In steady state, the solute transport rate from bulk solution to the membrane surface is equal to the reverse transport rate of solute to the bulk solution from the membrane surface plus the transport rate of solute pass thought the membrane with permeate (Fig. 5). The mass conservation at the boundary of concentration polarization region can be easily obtained [23] as

$$J(C' - C_p) = -D\frac{\mathrm{d}C'}{\mathrm{d}x} \tag{8}$$

where C' is the concentration of feed at the boundary layer, C_p is the concentration in permeate, D is the diffusion coefficient, and x is the distance from the membrane surface.

After integrating Eq. (8), a well-known equation can be obtained as

$$J = \frac{D}{\delta_p}\ln\left(\frac{C_m - C_p}{C_b - C_p}\right) \tag{9}$$

where δ_p is the boundary layer thickness, C_m is the concentration at the membrane surface, and C_b is the concentration of bulk solution.

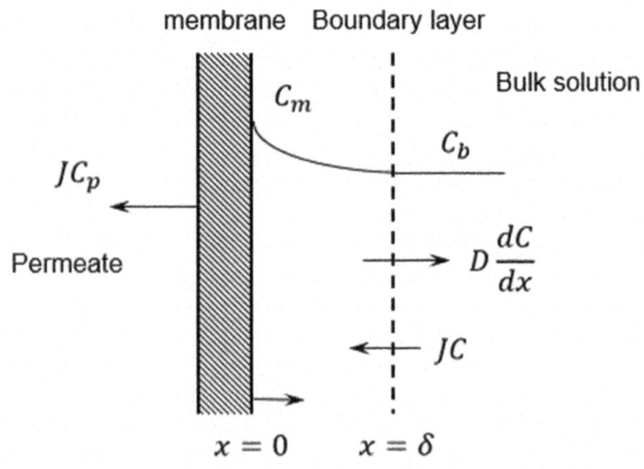

Fig. 5 Concentration polarization phenomena

Equation (9) is often used for the dead-end microfiltration process, while it will bring significant errors in the cross-flow microfiltration process. Therefore, the effective diffusion coefficient was used to replace the traditional diffusion coefficient in the formula (9) since the hydrodynamic effect plays a significant role in the cross-flow microfiltration process. For example, Zydeny et al. [24] considered the gradient of flow velocity at the membrane surface and characteristic parameters of feed and proposed a shear-induced diffusion coefficient to replace the general diffusion coefficient:

$$J(x) = 0.052\left(\frac{a}{x}\right)^{1/3} \gamma_w \ln\left(\frac{C_m}{C_b}\right) \tag{10}$$

where a is the particle radius, γ_w is the wall shear rate, and x is the axial coordinate along the channel length.

The length average flux can be obtained as

$$J = 0.078\left(\frac{a^4}{L_{\text{channel}}}\right)^{1/3} \gamma_w \ln\left(\frac{C_m}{C_b}\right) \tag{11}$$

Later, Belfort and others further improved and expanded the concentration polarization model application scope [25]. Finally, Davis and others [26] proposed the famous shear force-induced diffusion model after considering the effects of concentration and viscosity:

$$< J > = 0.072\left(\frac{\varphi_w w a^4}{\varphi_b L_{\text{channel}}}\right)^{1/3} \gamma_w \tag{12}$$

where φ_w and φ_b are the particle volume fraction in the boundary layer and the solids volume fraction in the suspension being filtered, respectively.

2.2 Individual (Single) Mechanism Model

The actual membrane pores, with different radii and lengths, are not convenient for us to analyze the fouling process and establish mathematical models. Therefore, for simplicity, the membrane pores are regarded as ideal parallel hollow cylinders with constant radius (r) and length (L).

The comparison of particle size and pore size of filtrate media is an extraordinary criterion for distinguishing different fouling mechanisms (Fig. 6). If particle sizes of foulants are bigger than the pore size of the membrane (filtrate media), a cake layer is formed. The cake layer can undertake a certain pressure drop between the feed and permeate sides, resulting in flux decay. Another situation is that the particle sizes are similar to the membrane's pore size (filtrate media), particles can block pores, and the decrease of open pore numbers leads to flux decay. In addition, if particle

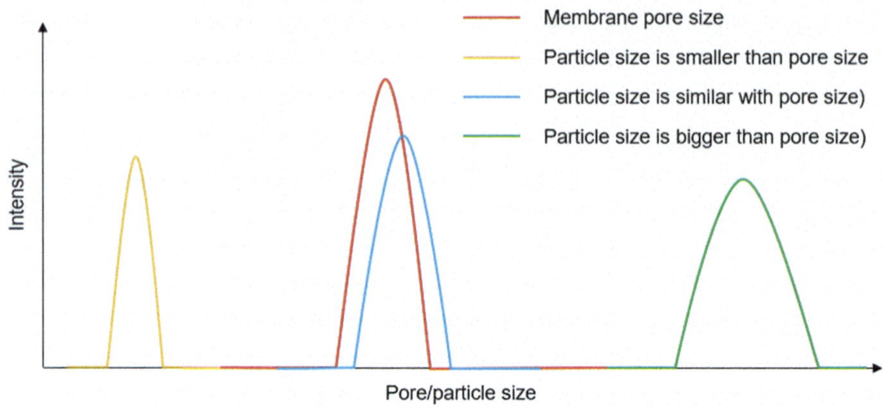

Fig. 6 Membrane pore size distribution and particle size distribution

sizes of foulants are smaller than the pore size of the membrane (filtrate media), the adsorption between the membrane (filtrate media) and particle, some particles have a chance to adsorb onto the pore walls, and it results in a pore radius decrease and flux decay alike. Until now, numerous types of research have been done to build a mathematical model for describing the filtrate volume, or the flux varies with time for this criterion of classifying the fouling mechanism.

2.2.1 Classical Pore-Blocking Law

Classical pore-blocking laws have provided the theoretical basis and played a significant role in model establishment for the membrane filtration process (fouling process). Therefore, figuring out this law is extremely important. The laws state that the fouling mechanism can be divided into four categories (Fig. 7): (1) **complete pore blocking**, (2) **standard pore blocking**, (3) **intermediate pore blocking**, and (4) **cake filtration**. Here, the formula for classical pore-blocking law under constant pressure mode is presented in Table 5 [27].

(a) **Complete pore blocking ($n = 2.0$)**

Particle sizes are almost identical to the membrane pore size for the complete pore-blocking mechanism. Thus, particles can completely block membrane pores, but from the probability perspective, the probability that a particle blocks an open pore is constant. It is easy to obtain a hypothesis: the number of blocked pores is directly proportional to the filtration volume (i.e., $N_0^{'} \propto v$). Then, the membrane filtration flux (J) is proportional to the number of open pores of the membrane, and it can be written as

$$J = \frac{dV}{dt} = k_c p N_0' = k_c p \left(N_0 - x' V\right) \tag{13}$$

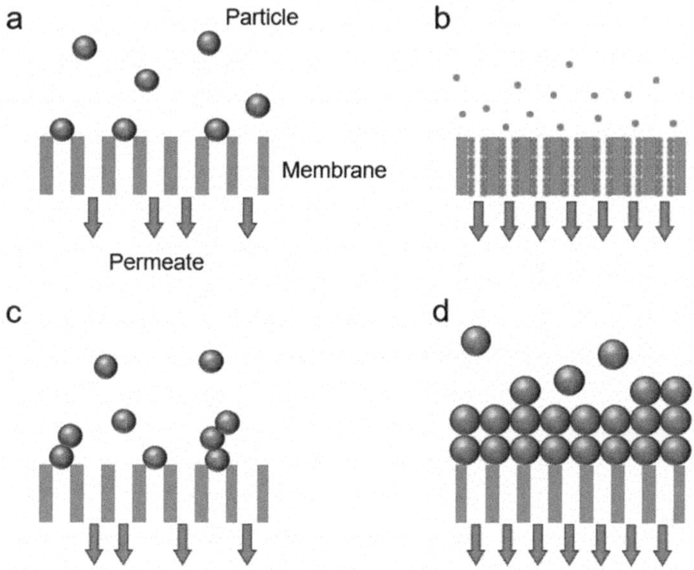

Fig. 7 Classical pore-blocking laws: **a** complete pore blocking, **b** standard pore blocking, **c** intermediate pore blocking, and **d** cake filtration

where V is the filtrate volume per unit effective membrane area, t is the filtration time, k_c is the proportional constant, p is the trans-membrane pressure drop, $N_0{}'$ is the number of open pores per unit effective membrane area at filtration time t, N_0 is the number of open pores per unit effective membrane area for a virgin membrane, and $x^{'}$ is the number of pores blocked per unit filtrate volume.

Here, the second derivative of t to v can be obtained according to Eq. (13)

$$\frac{d^2t}{dV^2} = k_c p x' \left(k_c p \left(N_0 - x'V\right)\right)^{-2} = K_b \left(\frac{dt}{dV}\right)^2 \tag{14}$$

where $K_b = k_c p x^{'}$.

It was evident that in a complete pore-blocking mechanism, the parameter n in $\frac{d^2t}{dV^2} = k\left(\frac{dt}{dV}\right)^n$ is equal to 2.0.

TERM I [$v = f(t)$]:

In complete pore-blocking mechanism, V as a function of t can be obtained by integrating Eq. (13) with the integral interval of $V = 0$ at $t = 0$ and $V = V$ at $t = t$.

$$V = \frac{N_0}{x'}(1 - \exp(-K_b t)) \tag{15}$$

In previous demonstrations, it is clear that when $t = 0$, $V = 0$, and the membrane filtration flux can be regarded as the initial filtration flux (J_0) (which is equal to

Table 5 The formulae of classical pore-blocking laws

Formulae	Complete pore blocking	Standard pore blocking	Intermediate pore blocking	Cake filtration
$\frac{d^2t}{dv^2} = k\left(\frac{dt}{dv}\right)^n$	$n = 2.0$	$n = 1.5$	$n = 1.0$	$n = 0$
$v = f(t)$	$v = \frac{J_0}{K_b}(1 - \exp(-K_b t))$	$\frac{t}{v} = \frac{K_s}{2}t + \frac{1}{J_0}$	$K_i v = \ln(1 + K_i J_0 t)$	$\frac{t}{v} = \frac{K_c}{2}t + \frac{1}{J_0}$
$J = f(t)$	$J = J_0 \exp(-K_b t)$	$J = J_0\left(\frac{K_s J_0}{2}t + 1\right)^{-2}$	$K_i t = \frac{1}{J} - \frac{1}{J_0}$	$J = \frac{J_0}{(1+2K_c J_0^2 t)^{1/2}}$
$J = f(v)$	$K_b v = J_0 - J$	$J = J_0\left(1 - \frac{K_s}{2}v\right)^2$	$J = J_0 \exp(-K_i v)$	$K_c v = \frac{1}{J} - \frac{1}{J_0}$

$k_c p N_0$). Substituting J_0 and K_b into Eq. (15) and the final expression of V can be obtained as

$$V = \frac{J_0}{K_b}(1 - \exp(-K_b t)) \tag{16}$$

TERM II [$J = f(t)$]:

In this term, J as a function of t can be easily obtained through the first derivative of V to t $\left(J = \frac{dV}{dt}\right)$ in Eq. (16)

$$J = J_0 \exp(-K_b t) \tag{17}$$

TERM III [$J = f(v)$]:

In this term, J as a function of V can be easily obtained by combining Eqs. (16) and (17)

$$K_b V = J_0 - J \tag{18}$$

(b) **Standard pore blocking (n = 1.5)**

For the standard pore-blocking mechanism, particle sizes are much smaller than the membrane pore size. Therefore, particles can enter the membrane pores and then adsorb onto the pore walls to reduce the pore radius. Here, to simplify this process, a hypothesis is achieved that the pore volume decreases proportionally to the filtrate volume per unit effective membrane area (V). One expression for the decreasing volume can be obtained as

$$\Delta V_{\text{pore}} = \frac{c}{1 - \varepsilon_p} dV \tag{19}$$

where c is the volume of particle trapped per unit filtrate volume (V) per unit membrane area, ε_p is the porosity of the particle layer formed on the membrane pore wall, and dv is the filtrate volume in a tiny time dt.

From a geometric approach, the decreased space of a single membrane pore can be unfolded into a cuboid (Fig. 8) and its volume can be expressed as

$$\Delta V_{\text{pore}} = -2\pi r L dr N_0 \tag{20}$$

where r is the membrane pore radius at time t, L is the length of the membrane pore, N_0 is the number of pores per unit membrane area, and dr is the variation of pore radius in a tiny time dt.

Hence, combining Eqs. (19) and (20) followed by defining the integral interval ($r = r_0$ at $v = 0$ and $r = r$ at $v = v$), the pore radius as a function of filtrate volume can be obtained after integrating

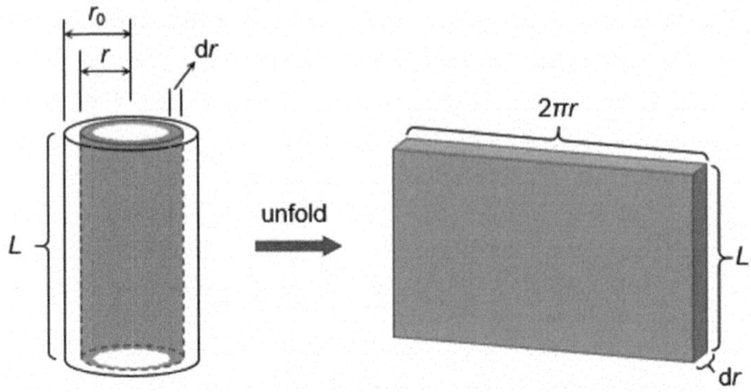

Fig. 8 The diagram of the space of foulant particles in one membrane pores and its unfolded drawing

$$\frac{r}{r_0} = \left(1 - \frac{c}{\pi L N_0 (1 - \varepsilon_p) r_0^2} V\right)^{1/2} = \left(1 - \frac{K_s}{2} V\right)^{1/2} \tag{21}$$

where K_s is the blocking constant for standard pore-blocking mechanism which is equal to $2c/(\pi L N_0(1 - \varepsilon_p) r_0{}^2)$.

Now supposing the membrane pores as capillaries. Hagen-Poiseuille law tells us that the flux across pores is proportional to the fourth power of the corresponding radius, and the relationship between flux and filtrate volume can be written as

$$\frac{J}{J_0} = \left(\frac{r}{r_0}\right)^4 = \left(1 - \frac{K_s}{2} V\right)^2 = \frac{dV}{J_0 dt} \tag{22}$$

Here, the second derivative of t to V for standard pore-blocking law can be obtained according to Eq. (22) and it is obvious that in complete pore-blocking mechanism, the parameter n in $\frac{d^2t}{dV^2} = k\left(\frac{dt}{dV}\right)^n$ is equal to 1.5.

$$\frac{d^2 t}{dV^2} = K_s J_0^{-1} \left(1 - \frac{K_s}{2} V\right)^{-3} = K_s J_0^{1/2} \left(J_0^{-1} \left(1 - \frac{K_s}{2} V\right)^{-2}\right)^{1.5} = K_s J_0^{1/2} \left(\frac{dt}{dV}\right)^{1.5} \tag{23}$$

TERM I [$V = f(t)$]:

In standard pore-blocking mechanism, V as a function of t can be obtained by integrating Eq. (22) with the integral interval of $V = 0$ at $t = 0$ and $V = V$ at $t = t$.

$$V = \frac{2 J_0 t}{K_s J_0 t + 2} \tag{24}$$

Rearranging and simplifying Eq. (24), the t/V as a function of t can be obtained as

$$\frac{t}{V} = \frac{K_s}{2}t + \frac{1}{J_0} \tag{25}$$

TERM II [$J = f(t)$]:

In this term, J as a function of t can be easily obtained through the first derivative of V to t in Eq. (24):

$$J = J_0\left(\frac{K_s J_0}{2}t + 1\right)^{-2} \tag{26}$$

TERM III [$J = f(V)$]:

In this term, J as a function of V can be easily obtained by combining Eqs. (25) and (26):

$$J = J_0\left(1 - \frac{K_s}{2}V\right)^2 \tag{27}$$

(c) **Intermediate pore blocking (*n* = 1.0)**

Particle sizes are almost identical to the membrane pore size for the intermediate pore-blocking mechanism. Thus, particles can block the membrane pores and complete the pore-blocking mechanism but differ in the probability that particles block open pores. The complete pore-blocking mechanism insists that the probability of particles blocking open pores is constant, while the intermediate pore-blocking mechanism emphasizes that this probability varies with filtrate volume. More filtrate volume can lead to fewer numbers of open pores, followed by a lower probability of foulant particles newly reaching the membrane surface and blocking pores. Instead, foulant particles prefer to deposit onto the other particles' previously blocked pores. Here, to describe the intermediate pore-blocking law, one hypothesis is made for intermediate pore blocking that the rate of pore blocking (($\mathrm{d}N_0'/\mathrm{d}V$) is proportional to the number of open pores at filtration time t

$$\frac{dN_0'}{dV} = -K_i N_0' \tag{28}$$

where K_i is the intermediate pore-blocking constant. The open pores at filtration time t can be easily obtained by integrating Eq. (28)

$$N_0' = N_0 \exp(-K_i V) \tag{29}$$

Substituting $N_0^{'}$ of Eq. (29) into Eq. (13) and the filtration flux can be written as

$$J = \frac{dV}{dt} = k_c p N_0 \exp(-K_i V) = J_0 \exp(-K_i V) \tag{30}$$

Thus, the second derivative of t to V can be obtained as follows and the parameter n in $\frac{d^2t}{dV^2} = k\left(\frac{dt}{dV}\right)^n$ for intermediate pore-blocking mechanism is obviously equal to 1.0.

$$\frac{d^2t}{dV^2} = \frac{K_i}{k_c p N_0} \exp(K_i V) = K_i \left(\frac{dt}{dV}\right)^1 \tag{31}$$

TERM I [$V = f(t)$]:

In intermediate pore-blocking mechanism, v as a function of t can be obtained by integrating Eq. (30) with the integral interval of $V = 0$ at $t = 0$ and $V = V$ at $t = t$.

$$K_i V = \ln(K_i J_0 t + 1) \tag{32}$$

TERM II [$J = f(t)$]:

In this term, J as a function of t can be easily obtained through the first derivative of v to t in Eq. (32)

$$J = \frac{J_0}{K_i J_0 t + 1} \tag{33}$$

Rearranging and simplifying Eq. (33) to obtain the final expression:

$$K_i t = \frac{1}{J} - \frac{1}{J_0} \tag{34}$$

TERM III [$J = f(V)$]:

In this term, J as a function of v can be easily obtained by combining Eqs. (32) and (33)

$$J = J_0 \exp(-K_i V) \tag{35}$$

(d) **Cake filtration ($n = 0$)**

For the cake filtration mechanism, particle sizes are bigger than the membrane pore size, and the cake is formed in the filtration process with no pores blocked. The cake is well studied in the industrial filtration process through the specific cake resistance

(α_{av}), the ratio of the mass of wet to the mass of dry cake (m). According to Ruth theory [28], the membrane filtration flux (J) for the filtration process with the cake is

$$J = \frac{dv}{dt} = \frac{p(1-\text{ms})}{\mu\alpha_{\text{av}}\rho s(V+V_m)} \tag{36}$$

where α_{av} is the average specific cake resistance, ρ is the filtrate density, s is the mass fraction of solids in colloids, m is the ratio of the mass of wet to the mass of dry cake, and V_m is the fictitious filtrate volume per unit membrane area required to obtain the cake with the flow resistance equivalent to that of the membrane.

According to Eq. (36), it is easy to get the second derivative of t to V for the cake filtration mechanism. The parameter n in $\frac{d^2t}{dV^2} = k\left(\frac{dt}{dV}\right)^n$ for the cake filtration is equal to zero:

$$\frac{d^2t}{dV^2} = \frac{\mu\alpha_{\text{av}}\rho s}{p(1-\text{ms})} = K_c\left(\frac{dt}{dV}\right)^0 \tag{37}$$

where K_c is the cake filtration constant which is equal to $\frac{\mu\alpha_{av}\rho s}{p(1-\text{ms})}$.

TERM I [$V = f(t)$]:

In cake filtration mechanism, v as a function of t can be obtained by integrating Eq. (36) with the integral interval of $V = 0$ at $t = 0$ and $V = V$ at $t = t$.

$$\frac{K_c}{2}V^2 + K_c V_m V = t \tag{38}$$

$$V = \frac{-K_c V_m + \sqrt{(K_c V_m)^2 + 2K_c t}}{K_c} \tag{39}$$

Rearranging Eq. (38) to obtain

$$\frac{t}{V} = \frac{K_c}{2}V + K_c V_m \tag{40}$$

According to Eq. (36), when filtration time is 0, as well as filtrate volume v, and the expression of the initial filtration flux (J_0) is

$$J_0 = \left(\frac{\mu\alpha_{av}\rho s}{p(1-ms)}V_m\right)^{-1} = (K_c V_m)^{-1} \tag{41}$$

Thus, Eq. (38) can be rewritten as

$$\frac{t}{V} = \frac{K_c}{2}V + \frac{1}{J_0} \tag{42}$$

TERM II [$J = f(t)$]:

In this term, J as a function of t can be easily obtained through the first derivative of v to t in Eq. (39)

$$\frac{J}{J_0} = \left(1 + 2K_c J_0^2 t\right)^{-0.5} \tag{43}$$

TERM III [$J = f(V)$]:

In this term, J as a function of V can be easily obtained by substituting Eq. (38) into Eq. (43)

$$J = J_0(1 + K_c J_0 V)^{-1} \tag{44}$$

Rearranging Eq. (44) to obtain

$$K_c V = \frac{1}{J} - \frac{1}{J_0} \tag{45}$$

2.2.2 Cake Thickness Model

Different operations can bring different fouling behaviors for the cake filtration mechanism, especially for stirring operations in lab work. Numerous types of research have been done in this field. In 1992, Wang and co-workers established a mathematical model, with an index Φ firstly proposed, to describe the variation of the membrane flux [29]. In Wang's model, the flux is determined according to Darcy's law:

$$J = \frac{\Delta P}{\mu[R_m + r_c \delta_c]} \tag{46}$$

where r_c is the specific cake resistance (m^{-2}) and δ_c is the cake thickness (m).

Here, the cake thickness can be written as

$$\delta_c = \delta_0 \int_0^t J dt \tag{47}$$

where δ_0 is the volume of cake formed on the surface of the membrane to get unit volume of permeate (m^3/m^3).

Substituting Eq. (44) into Eq. (46):

$$J = \frac{\Delta P}{\mu\left[R_m + r_c\delta_0 \int_0^t J dt\right]} \tag{48}$$

Assuming $X = \int_0^t J dt$, Eq. (48) can be rewritten as

$$\frac{dX}{dt} = \frac{\Delta P}{\mu[R_m + r_c\delta_0 X]} \tag{49}$$

Integrating it from $t = 0\ X = 0$ to $t = t\ X = X$:

$$\frac{1}{2} r_c\delta_0 X^2 + R_m X - \frac{\Delta P}{\mu} t = 0 \tag{50}$$

$$X = \frac{-R_m + \sqrt{R_m^2 + 2r_c\delta_0 \Delta P t/\mu}}{r_c\delta_0} \tag{51}$$

Thus, J as a function of time can be solved

$$J = \frac{J_0}{\sqrt{1 + 2\frac{r_c\delta_0}{R_m} J_0 t}} = \frac{J_0}{\sqrt{1 + 2\Phi J_0 t}} \tag{52}$$

Wang's model can predict the flux decay in the microfiltration processes. However, some issues still exist. Z. Wang et al. mentioned that the reasons for the deviations in experimental data and predicted value for Wang's model were three factors [30, 31]. (1) In the initial filtration stage, no cake layer formed. (2) Different interactions can be observed between different foulants and the membrane. And (3) the cake layer structure was thought to be varied. Based on these considerations, Wang and co-workers reinvestigated the index Φ and introduced another parameter ($\overline{k}$) to correct the prediction.

$$J = \frac{J_0}{\sqrt{1 + 2\overline{k}\Phi J_0 t}} \tag{53}$$

This semi-empirical flux model (Eq. (53)) can be used for sewage, soy sauce, or other suspensions. According to the experimental data, the filtration flux curves change with time under certain pressure and then get the value of J_0 and Φ value (average value) through regression. The final semi-empirical model can be obtained by putting the above Φ value and $\overline{k}$ value into formulae.

Table 6 shows the corresponding parameters of this semi-empirical model under some conditions. It can be seen that under different operating pressures, the correction coefficient $\overline{k}$ is between 2.5 and 1.0, and with the decrease of pressure from 0.20 to 0.08 MPa, the value of $\overline{k}$ tends to be close to 1. For example, $\Delta P = 0.20$ MPa, $k = 2.317$; when $\Delta P = 0.08$ MPa, $k = 1.293$.

Table 6 Model parameter in reference [30]

Δp/MPa	$\Phi \times 10^{-3}$	$\overline{k}$	Model prediction
0.20	1.475	2.317	$J_0/\sqrt{1+2\times 2.317\times 1.475\times 10^3\times 3.992\times 10^{-6}t}$
0.18	1.535	1.824	$J_0/\sqrt{1+2\times 1.823\times 1.535\times 10^3\times 3.862\times 10^{-6}t}$
0.16	1.659	1.767	$J_0/\sqrt{1+2\times 1.767\times 1.659\times 10^3\times 4.767\times 10^{-6}t}$
0.12	1.377	1.678	$J_0/\sqrt{1+2\times 1.678\times 1.377\times 10^3\times 3.470\times 10^{-6}t}$
0.10	1.151	1.622	$J_0/\sqrt{1+2\times 1.622\times 1.151\times 10^3\times 3.315\times 10^{-6}t}$
0.08	1.897	1.293	$J_0/\sqrt{1+2\times 1.293\times 1.897\times 10^3\times 2.251\times 10^{-6}t}$

The model parameter has changed nonlinearly with operating conditions, and its time-dependent character can result in a considerable error in the model prediction. Therefore, finding a suitable model parameter for the membrane flux prediction model was also assessed [32]. The higher predictive accuracy model can be obtained after the constant membrane fouling index in the model is replaced by the average membrane fouling index or a composite parameter with a free combination of the initial membrane fouling index, meantime minimum, and the maximum of the model parameter.

These proposed models above were analyzed based on the cake thickness for the situation without any shear stress. Silva and co-workers proposed models to predict the variation of the membrane flux with/without shear force in the filtration process (Fig. 9) [33]. Based on the mass balance, Silva established the proper mass balance equations for dead-end filtration mode (without shear force (Eq. (54)) and cross-flow filtration mode (with shear force (Eq. (55)).

$$\left(J+\frac{d\delta_c}{dt}\right)\cdot\phi_b=\phi_c\cdot\frac{d\delta_c}{dt} \tag{54}$$

$$\left(J+\frac{d\delta_c}{dt}\right)\cdot\phi_b-k_1\left(\frac{dU}{dy}\right)\delta_c=\phi_c\cdot\frac{d\delta_c}{dt} \tag{55}$$

where φ_b is the solid volume fractions in suspension, φ_c is the solid volume fractions in cake, U is the liquid velocity, k_1 is a constant, and dU/dy means the velocity gradient.

If focus on dead-end filtration mode without agitation (without shear force), i.e., Eq. (54). It can be rewritten as

$$\delta_c=\frac{\phi_b}{\phi_c-\phi_b}\int_0^t Jdt \tag{56}$$

Substituting Eq. (56) into Eq. (46) and the final expression for dead-end filtration without agitation (the case without shear force), we can obtain

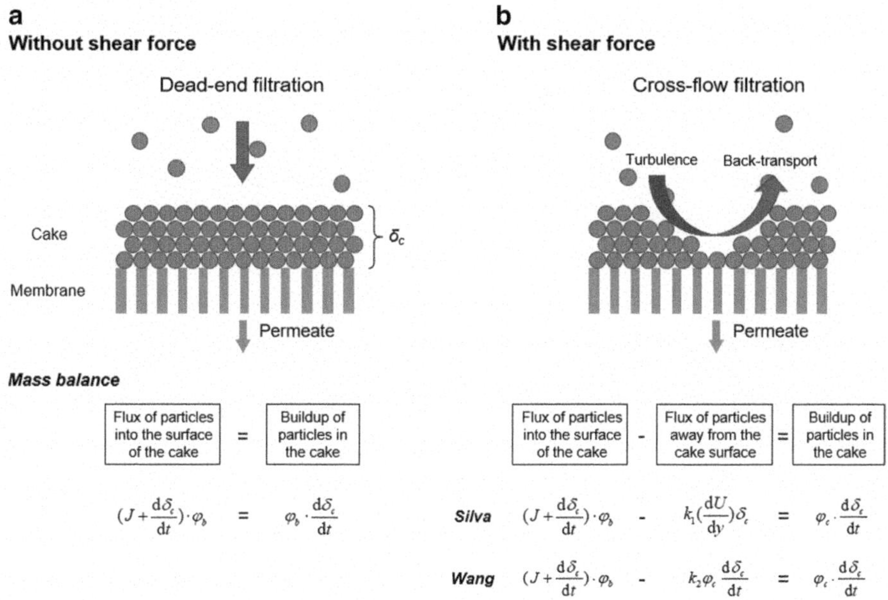

Fig. 9 Diagram of the cake formation in dead-end filtration case (**a**) and cross-flow filtration case (**b**)

$$J = \frac{J_0}{\sqrt{1 + \frac{2r_c \Delta P}{\mu R_m^2}\frac{\phi_b}{\phi_c - \phi_b}t}} \tag{57}$$

In the case for cross-flow filtration (with shear force), back-transport should be considered and this term can be expressed as $k_1\left(\frac{dU}{dy}\right)\delta_c$. To solve and obtain the final expression, Eq. (55) should be rearranged:

$$\frac{d\delta_c}{dt} = \frac{\Delta P}{\mu r_c \delta_c}\frac{\phi_b}{\phi_c - \phi_b} - k_1\left(\frac{dU}{dy}\right)\frac{\delta_c}{\phi_c - \phi_b} \tag{58}$$

After integrating the cake layer thickness under high-shear case, the following expression can be obtained as

$$\delta_c = \sqrt{\frac{\Delta P \phi_b}{r_c k_1 \mu (dU/dy)}\left(1 - \exp(\frac{-2k_1(dU/dy)t}{\phi_c - \phi_b})\right)} \tag{59}$$

Thus, the filtration flux can be obtained as

$$J = \frac{\Delta P}{\sqrt{\frac{\Delta P \phi_b \mu r_c}{k_1(dU/dy)}\left(1 - \exp(\frac{-2k_1(dU/dy)t}{\phi_c - \phi_b})\right)} + \mu R_m} \tag{60}$$

Here, for the Newtonian fluid, the velocity gradient dU/dy can be obtained as

$$\frac{dU}{dt} = \frac{\tau_w}{\mu_1} \tag{61}$$

where τ_w is the shear stress and μ_1 is the suspension viscosity.

In the meantime, the compressibility of the cake layer was employed by using:

$$r_c = r_0(\Delta P)^n \rho_s \phi_c \tag{62}$$

where s is the cake compressibility and ρ_s is the mass density of solids comprising the cake.

The membrane flux of the Silva model under high-shear case can be obtained by substituting Eqs. (61) and (62) into Eq. (60):

$$J = \frac{\Delta P}{\sqrt{\frac{\Delta P^{1+n}\phi_b \mu \mu_1 r_0 \rho_s \phi_c}{k_1 \tau_w}\left(1 - \exp\left(\frac{-2k_1 \tau_w t}{\mu_1(\phi_c - \phi_b)}\right)\right) + \mu R_m}} \tag{63}$$

In 2007, Z. Wang, J. S. Chu, et al. also established a mathematical model describing the cake filtration of stirred dead-end microfiltration mode [7]. Wang and co-workers considered and rewrote the back-transport term as $k_2\phi_c\frac{d\delta_c}{dt}$ (Fig. 9). Similarly considering the cake compressibility, the final expression of Wang's model is

$$J = \frac{\Delta P}{\sqrt{\frac{2r_0 \Delta P^{1+n}\phi_b \mu}{\phi_c(k_2+1) - \phi_b} t + \mu^2 R_m^2}} \tag{64}$$

where k_2 is the denudation coefficient.

In order to consider the geometric parameters of module and feed properties on the shear stress, Z. Wang and Z.F. Gao rewrote the plate and frame module model for cross-flow mode [34]. They corrected the shear stress in the cross-flow situation as

$$\tau_w = \frac{2h_m \overline{u} \Delta P}{L' \mu_1} \tag{65}$$

where $\overline{u}$ is the average flow velocity, h_m is the channel height of filtration module, and L' is the length of the channel of filtration module.

Therefore, the corrected model is

$$J = \frac{\Delta P}{\sqrt{\frac{\Delta P^n \phi_b \mu \mu_1 r_0 \rho_s \phi_c}{2k_1 u h_m}\left(1 - \exp\left(\frac{-4k_1 \Delta P h_m u t}{L' \mu_1(\phi_c - \phi_b)}\right)\right) + \mu R_m}} \tag{66}$$

2.2.3 Corrected Cake Resistance by Wang

Different foulants have related properties, e.g., surface charge and particle size. Besides, the cake formed by these foulants is compressible due to its properties. Thus, it is crucial to have an assessment criterion to describe the fouling behaviors for various foulants. For this purpose, Wang and co-workers established a mathematical cake filtration model and simultaneously considered the deposition properties of different foulant particles and the cake compressibility [35].

From cake filtration law, it is evident that the t/V and V satisfy the relationship below:

$$\frac{t}{V} = \frac{\alpha \mu C_b}{2 \Delta P A^2} V + \frac{1}{J_0 A} \quad (67)$$

where A is the membrane area, m^2.

Meanwhile, it is easy to obtain the cake resistance:

$$R_c = \frac{\alpha M}{A} = \frac{\alpha C V}{A} \quad (68)$$

where M is the cake mass, kg.

Considering the cake compressibility, i.e., $\alpha = \alpha_0(\Delta P/P_0)^n$, Eq. (68) can be rewritten as

$$R_c = \frac{\beta (\Delta P)^n C_b V}{A} \quad (69)$$

where β is a constant (m·kg^{-1}·Pa^{-n}) and n is the compressibility constant.

Similarly, the cake filtration expression considering cake compressibility can also be obtained by substituting $\alpha = \alpha_0(\Delta P/P_0)^n$ into Eq. (67):

$$\frac{t}{V} = \frac{\beta (\Delta P)^{n-1} \mu C_b}{2 A^2} V + \frac{1}{J_0 A} \quad (70)$$

Another factor, i.e., the deposition properties of different foulants, should be considered. In the previous studies, the cake mass was considered linearly increased with the bulk concentration [27]. However, Wang noted that due to the back-transport, convective, and diffusion mechanisms, the cake mass did not linearly increase with the bulk concentration but followed the power law:

$$m = C^{k'} V \quad (71)$$

where k' is the deposition factor.

Thus, the corrected cake filtration law can be achieved by combining Eqs. (70) and (71):

$$\frac{t}{V} = \frac{\beta^*(\Delta P)^{n-1}\mu C_b^{k'}}{2A^2}V + \frac{1}{J_0 A} \tag{72}$$

where β^* is the corrected parameter β. Comparing Eqs. (70) and (72), it is easily to obtain the relationship between β^* and β:

$$\beta = \beta^* C_b^{k'-1} \tag{73}$$

Here, a corrected cake model considering the deposition properties of different foulants and the cake compressibility is wholly achieved. However, this cake model is only considered the whole property of a cake. Some local properties, like the local porosity and specific cake resistance, are not considered. Therefore, a new cake model must be established to describe these properties. Fortunately, Zhu and co-workers have overcome these issues and established their model to describe the variations of whole properties and the local properties in the filtration process.

2.2.4 Traditional Filtration Theory in Membrane Process

The traditional industrial filtration process for solid/liquid separation can be divided into cake and deep bed filtration. Cake filtration refers that the solid particles being rejected on the filtration media surface. In contrast, deep bed filtration refers to these solids being deposited into the filtration media to achieve the separation goal [36]. From the definition of these concepts, it is apparent that cake filtration in the traditional filtration process is similar to the so-called cake filtration in pore-blocking law, and deep bed filtration is similar to standard pore blocking in pore-blocking law. However, one thing should be noticed traditional filtration theory was proposed much earlier than pore-blocking law; how to draw lessons from traditional filtration theory to build up an accurate mathematical description of the membrane process is a vital issue.

The research on cake filtration in the traditional process can be traced back to the early twentieth century. B. F. Ruth gave a perfect derivation process for cake filtration in 1935 [28]. In 1953, F. M. Tiller published a paper about the research on the porosity in the filtration process [37], and a series of works were done during the next few decades [38–49]. In 2001, Tien summarized the related work and listed four relationships between solid compressive stress and liquid pressure [50]. To apply traditional filtration theory in the membrane process, Z. Zhu and co-workers combined it with membrane filtration theory, proposed a mathematical model, and gave an excellent prediction for the membrane fouling process [51].

(a) **Cake filtration in traditional filtration theory**

In 1935, Ruth proposed that the flow rate was proportional to the driving pressure and the reciprocal of the resistance when fluid flowed across a porous medium. This relationship can be expressed as

$$\frac{\mu}{A}\frac{dV}{dt} = \frac{p}{R} \tag{74}$$

where μ is the liquid viscosity, A is the filtration medium area, V is the permeate volume, t is the filtration time, p is the applied pressure, and R is the filtration resistance.

Here, the resistance includes the pipe resistance of the filtration system, the filtration media resistance, the cake resistance, and so on. Fortunately, the resistance-in-serials model is still available here and the detailed resistance can be divided into

$$R = R_1' + R_1'' + R_2 + R_3 + R_4 + \ldots \tag{75}$$

where R_1' is the separation media resistance, R_1'' is the resistance of any throttling layer, R_2 is the cake resistance, R_3 is the pipe resistance of the filtration system, and R_4 is the resistance of any valves.

Generally speaking, R_3 and R_4 could be ignored, R_1' and R_1'' are regarded as constants. Here, a relationship between the resistance and cake mass can be easily obtained as follows:

$$R = (w + w_c)\frac{\alpha}{A} \tag{76}$$

where w is the cake mass, w_c is the virtual mass, and α is the specific cake resistance.

It should be emphasized that the virtual mass w_c is equal to the cake mass that has the same resistance value as filter media. Meanwhile, another thing that should be noticed is the mass w here do not refer to the weight of the cake per unit area. Hence, to calculate the cake resistance, mass should divide by the filtration area.

It is necessary to convert w into easy-to-measure parameters to adapt for the industrial application. Ruth defined the following equations:

$$m = \frac{\text{particle} + \text{water}_{\text{cake}}}{\text{particle}} \tag{77}$$

$$s = \frac{\text{particle}}{\text{particle} + \text{water}} = \frac{\text{particle}}{\text{particle} + \text{water}_{\text{cake}} + \text{water}_{\text{permeate}}} \tag{78}$$

Here, m refers to the wet-to-dry cake mass ratio and s refers to the particle mass fraction of feed. Therefore, w can be written as

$$w = \frac{V\rho s}{1 - \text{ms}} \tag{79}$$

And Eq. (76) can be rewritten as

$$R = (V + V_m)\frac{\rho s}{1 - \text{ms}}\frac{\alpha}{A} \tag{80}$$

where V is the permeate volume and V_m is the virtual permeate volume.

Earlier research pointed out that some cakes were not rigid but could be compressed under high applied pressure. As a result, their specific cake resistance was linearly increased with the applied pressure. However, numerous experiments demonstrated that their specific cake resistance increasing trend was not linear. Thus, Ruth proposed a corrected specific cake resistance expression:

$$\alpha = \alpha_0\left(1 + \beta' p^n\right) \tag{81}$$

where α_0, $\beta^{'}$, and n are coefficients.

Therefore, substituting Eqs. (80) and (81) into Eq. (74) to obtain the modified relationship between the filtration flux and operation pressure:

$$\frac{\mu}{A}\frac{dV}{dt} = \frac{p}{\alpha_0(1 + \beta' p^n)}\frac{(1 - \text{ms})A}{(V + V_m)\rho s} \tag{82}$$

Owing to the pressure drop exists spatial variation, specific cake resistance also varies with cake structure spatial location. Rearranging Eq. (82):

$$\frac{p}{\alpha_0(1 + \beta' p^n)} = \frac{\mu}{A}\frac{dV}{dt}\frac{(V + V_m)\rho s}{(1 - \text{ms})A} = j \tag{83}$$

So, if the resistance of different regions of cake needs to be calculated, differential equation should be written as

$$\frac{dp}{dj} = \frac{\alpha_0(1 + \beta' p^n)^2}{1 + (1 - n)\beta' p^n} = \alpha_p \tag{84}$$

and the specific cake resistance for each region can be expressed as follows:

$$(dw/A)\alpha_p \tag{85}$$

In the next few decades, F. M. Tiller, H. R. Cooper, and M. Shirato, et al. had done detailed research on the cake porosity and compressive pressure.

Based on Ruth equation, Tiller and co-workers proposed a basic equation to describe fluid flow across porous media:

$$-g_c\frac{dp_x}{dw_x} = \alpha_x \mu q \tag{86}$$

where g_c is the conversion factors, the subscript x is the position information about the cake layer, i.e., the cake layer height information.

Here, the cake mass at the position x (w_x) could be written as follows:

$$dw_x = \rho_s \varepsilon_s dx = \rho_s(1 - \varepsilon)dx \tag{87}$$

To obtain the relationship between the compressive pressure and the spatial information, Eq. (87) was substituted into Eq. (86). Defining the lower limit of integral as 0 and the upper limit as x or l, respectively. Thus, relation could be obtained as

$$\int_0^x dx = x = \frac{g_c}{\mu \rho_s q} \int_0^{p_s} \frac{dp_s}{\alpha_x (1 - \varepsilon)} \tag{88}$$

$$\int_0^l dx = l = \frac{g_c}{\mu \rho_s q} \int_0^{p} \frac{dp_s}{\alpha_x (1 - \varepsilon)} \tag{89}$$

By comparing x and l, the variation of compressive pressure and cake relative spatial location of cake was

$$\frac{x}{l} = \frac{\int_0^{p_s} \frac{dp_s}{\alpha_x (1-\varepsilon)}}{\int_0^{p} \frac{dp_s}{\alpha_x (1-\varepsilon)}} \tag{90}$$

This information is derived from a spatiotemporal variation of numerous cake properties (e.g., porosity, permeability, specific cake resistance, compressive pressure). Some detailed derivations are shown in the next episode.

(b) **Zhu's model**

In the traditional industrial filtration process, as discussed above, the filter cake is studied a lot. It is related to the pressure across itself, and some local properties are different, like the specific cake resistance, the porosity, and the permeability. Tiller's power law is proved to have a good description of filter cake in the whole cake properties and the local cake properties. However, determining the parameters of Tiller's power law in the membrane fouling process is one of the burning issues. Zhu and co-workers determined the parameters and established a new model based on Tiller's power law to describe the properties of filter cake in the membrane process and the flux decay [50].

Derivation of Zhu's model

Before the derivation, it should be emphasized that the total pressure drop (Δp) is divided into (a) the pressure drop across the membrane (Δp_m) and (b) the pressure drop across the filter cake (Δp_c) (Fig. 10).

$$\Delta p = \Delta p_m + \Delta p_c \tag{91}$$

In terms of the local compressible filter cake properties (e.g., the local porosity, the specific cake resistance, and the permeability), these properties and the internal pressure satisfy the equations below according to Tiller's power law:

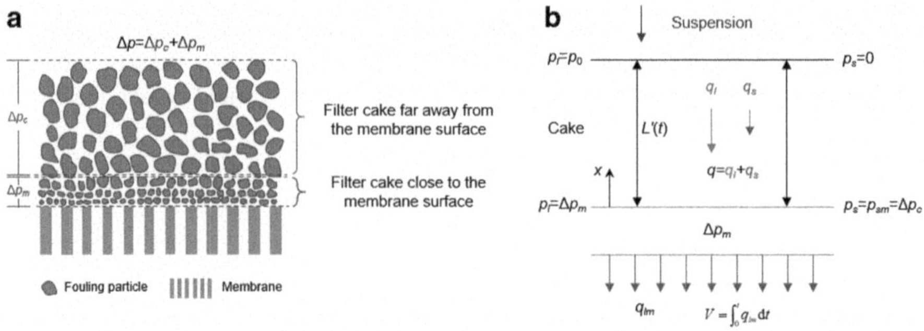

Fig. 10 **a** Diagram of the compressible filter cake and **b** the corresponding parameters along the cake

$$\varepsilon_s = \varepsilon_s^0 \left(1 + \frac{p_s}{p_a}\right)^{\beta} \tag{92}$$

$$k = k^0 \left(1 + \frac{p_s}{p_a}\right)^{-\delta} \tag{93}$$

$$\alpha = \alpha^0 \left(1 + \frac{p_s}{p_a}\right)^{n} \tag{94}$$

where ε_s is the volume fraction of particle; k is the cake permeability; α is the specific cake resistance; superscript 0 means the corresponding value when the compressible pressure is 0; p_s is the solid compressive pressure; p_a is the normalizing parameter of compressive pressure; β, n, and δ are empirical constant. The relationship among β, n, and δ can be expressed as

$$\alpha = 1/(\varepsilon_s k \rho_s) \tag{95}$$

$$\delta = n + \beta \tag{96}$$

Meanwhile, the total pressure drop can be written as another form:

$$\Delta p = p_l + p_s \tag{97}$$

where p_l is the pressure the liquid in cake undertaken and p_s is the pressure the solid in cake undertaken. If we take presume at the level of the membrane surface is zero, the cake growth direction is the positive direction (x-axis), when $x = 0$ $p_s = \Delta p_c$ and $x = l$ $p_s = 0$ (Fig. 10b). Here, l is the cake thickness and the differential of Eq. (97) is $\mathrm{d}p_l + \mathrm{d}p_s = 0$.

Here, the apparent rate of filtration consists of liquid apparent rate (q_l) and solid apparent rate (q_s). Assuming the solid apparent rate can be ignored, it can be written as

$$q = q_l = \frac{1}{A}\frac{dV}{dt} \tag{98}$$

Another form about the apparent rate of filtration is

$$\frac{q_l}{\varepsilon} - \frac{q_s}{\varepsilon_s} = -\frac{1}{\varepsilon}\frac{k}{\mu}\frac{\partial p_l}{\partial x} \tag{99}$$

Integrating Eq. (99) from $x = 0$ to $x = x$ and from $x = 0$ to $x = l$ to obtain:

$$\mu q x = \frac{k^0 p_a}{1-\delta}\left[(1+\frac{p_s}{p_a})^{1-\delta} - (1+\frac{\Delta p_c}{p_a})^{1-\delta}\right] \tag{100}$$

$$\mu q l = \frac{k^0 p_a}{1-\delta}\left[1 - (1+\frac{\Delta p_c}{p_a})^{1-\delta}\right] \tag{101}$$

Hence, the information of the position x/l can be obtained after Eq. (101) is divided by Eq. (100):

$$\frac{x}{l} = 1 - \frac{(1+\frac{p_s}{p_a})^{1-\delta} - 1}{(1+\frac{\Delta p_c}{p_a})^{1-\delta} - 1} \tag{102}$$

Rearranging Eq. (102) to achieve the p_s as a function of x/l:

$$\frac{p_s}{p_a} = \left[1 + \left(1 - \frac{x}{l}\right)\left(\left(1 + \frac{\Delta p_c}{p_a}\right)^{1-\delta} - 1\right)\right]^{\frac{1}{1-\delta}} - 1 \tag{103}$$

Meanwhile, p_l as a function of x/l can also be obtained by substituting Ep. (103) into Eq. (97):

$$\frac{p_l}{p_a} = \frac{\Delta p}{p_a} - \left[1 + \left(1 - \frac{x}{l}\right)\left(\left(1 + \frac{\Delta p_c}{p_a}\right)^{1-\delta} - 1\right)\right]^{\frac{1}{1-\delta}} + 1 \tag{104}$$

Up to now, the local compressible cake properties can be rewritten as

$$\varepsilon_s = \varepsilon_s^0\left[\left(\left(1 + \frac{\Delta p_c}{p_a}\right)^{1-\delta} - 1\right)\left(1 - \frac{x}{l}\right) + 1\right]^{\frac{\beta}{1-\delta}} \tag{105}$$

$$k = k^0\left[\left(\left(1 + \frac{\Delta p_c}{p_a}\right)^{1-\delta} - 1\right)\left(1 - \frac{x}{l}\right) + 1\right]^{\frac{-\delta}{1-\delta}} \tag{106}$$

$$\alpha = \alpha^0 \left[\left(\left(1 + \frac{\Delta p_c}{p_a} \right)^{1-\delta} - 1 \right) \left(1 - \frac{x}{l} \right) + 1 \right]^{\frac{n}{1-\delta}} \tag{107}$$

Thus, the cake layer resistance as a function of position x/l can be achieved based on the relationship between the specific cake resistance and the cake mass:

$$R_x = \int_0^x \alpha dw = \rho_s \int_0^x \alpha \varepsilon_s dx \tag{108}$$

$$R_x = -\rho_s \varepsilon_s^0 \alpha^0 (1-\delta) \frac{l}{C''} \left(\left[C''(1 - x/l) + 1 \right]^{\frac{1}{1-\delta}} - (1 + C'')^{\frac{1}{1-\delta}} \right) \tag{109}$$

$$R_c = -\rho_s \varepsilon_s^0 \alpha^0 (1-\delta) \frac{l}{C''} \left[1 - (1 + C'')^{\frac{1}{1-\delta}} \right] \tag{110}$$

where R_x is the cake resistance above the membrane surface of distance x, w is the cake mass from the membrane surface to the distance x, ρ_s is the cake density, and C is a parameter which is used to simplify formulae $C'' = (1+\Delta p_c/p_a)^{1-\delta} - 1$.

Substituting Eq. (110) into Darcy's law ($J = \Delta p/(R_m + R_c)$):

$$J = \frac{\Delta p}{\mu \left(R_m - \rho_s \varepsilon_s^0 \alpha^0 (1-\delta) \frac{l}{C''} \left(1 - (1 + C'')^{\frac{1}{1-\delta}} \right) \right)} \tag{111}$$

Now, only one problem should be solved here: how to determine the cake layer thickness L' as a function of filtration time. According to the mass balance, the increased thickness of the cake layer dl in a short time dt can be written as

$$\frac{dl}{dt} = \frac{-q_{l|l^+} - (-q_{l|l^-})}{\varepsilon_{l|l^+} - \varepsilon_{l|l^-}} \tag{112}$$

where L^+ is the position that has a finite distance to l whose direction is positive direction and L^- is the position that has a finite distance to l whose direction is negative direction. $\varepsilon_{l|l^-} = \varepsilon_s^0$ and $\varepsilon_{l|l^+} = \varepsilon_{s0} = s$. Here, s is the particle volume fraction in the bulk suspension.

As discussed above, the total apparent rate of filtration can be divided into two terms: q_l and q_s. For $x = l^+$ and $x = l^-$:

$$q_{l|l^-} + q_{s|l^-} = q_{l|l^+} + q_{s|l^+} = \left(-\frac{k}{\mu} \frac{\partial p_l}{\partial x} \right)_{x=0} \tag{113}$$

Combining Eqs. (99) and (113) to obtain:

$$q_{l|l^-} = \varepsilon_s^0 \left(-\frac{k}{\mu} \frac{\partial p_l}{\partial x} \right)_{x=l^-} + \left(1 - \varepsilon_s^0 \right) \left(-\frac{k}{\mu} \frac{\partial p_l}{\partial x} \right)_{x=0} \tag{114}$$

$$q_{l|l^+} = (1-\varepsilon_s^0)(-\frac{k}{\mu}\frac{\partial p_l}{\partial x})_{x=0} \tag{115}$$

Here, substituting Eqs. (114) and (115) into Eq. (112) and the variation of the cake thickness with filtration time can be rewritten as

$$\frac{dl}{dt} = \frac{k}{\mu}\left(\frac{\varepsilon_s^0}{\varepsilon_s^0 - s}\left(\frac{\partial p_l}{\partial x}\right)_{x=l} - \left(\frac{\partial p_l}{\partial x}\right)_{x=0}\right) \tag{116}$$

Now, it is easily to obtain the partial differential of p_l to x from Eq. (104):

$$\frac{\partial p_l}{\partial x} = \frac{p_a}{1-\delta}\frac{C''}{l}\left[1 + C''\left(1 - \frac{x}{l}\right)\right]^{\frac{\delta}{1-\delta}} \tag{117}$$

Substituting $x = 0$ and $x = l$ into Eq. (117), followed by combining Eq. (116) to obtain a first-order differential equation:

$$\frac{dl}{dt} = \frac{s}{\varepsilon_s^0 - s}\frac{k^0}{\mu}\frac{p_a}{1-\delta}\frac{C''}{l} \tag{118}$$

Finally, the cake thickness as a function of time can be achieved by integrating Eq. (118) from $t = 0\ l = 0$ to $t = t\ l = l$.

$$l = \sqrt{\frac{2S}{\varepsilon_s^0 - s}\frac{k^0}{\mu}\frac{p_a}{1-\delta}\int_0^t\left[\left(1 + \frac{\Delta p_c}{p_a}\right)^{1-\delta} - 1\right]dt} \tag{119}$$

Application of Zhu's model

Zhu's model presented the application of traditional filtration theory in the membrane process. It exhibited an excellent result in predicting the membrane flux variation (Fig. 11) and detailed spatiotemporal variation about cake properties in the membrane process.

Zhu's model demonstrated that the cake layer consisted of a skin layer and a loose layer. The skin layer was below the loose layer which was at location of $x/l = 0.2$, but contained over 70% resistance (Fig. 12).

Meanwhile, Zhu's model also demonstrated that other cake properties also varied sharply until $x/l = 0.2$ (Fig. 13).

2.2.5 Kelly-Zydney Model [52]

Protein fouling is one of the severe factors in the microfiltration process. In 1995, Kelly and Zydney proposed a mathematical model to describe the dual-mode fouling

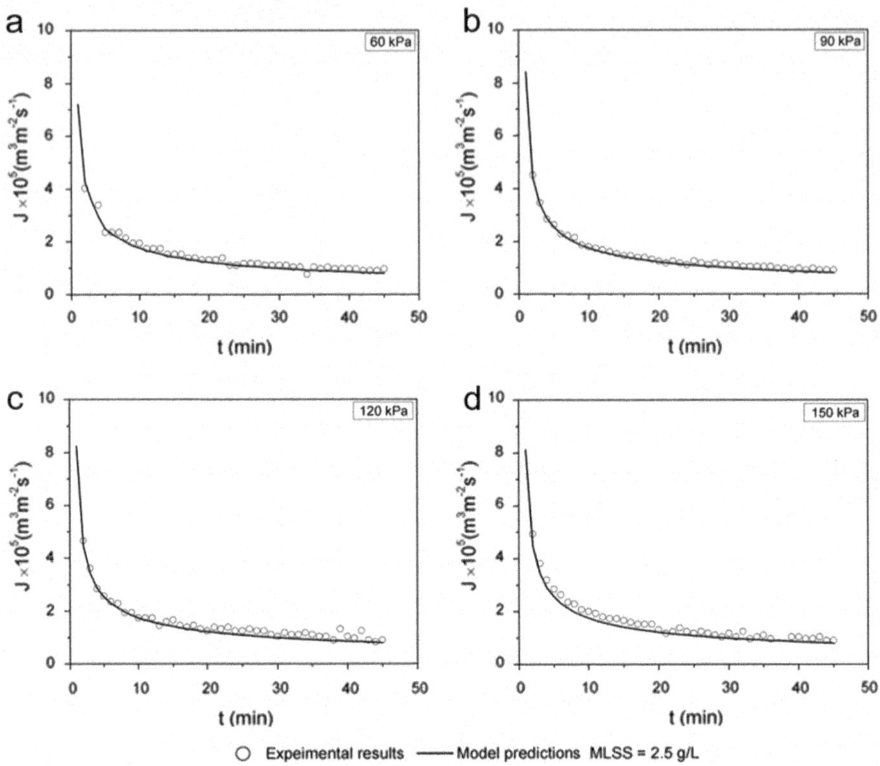

Fig. 11 Predictions of Zhu's model under different operation pressures

process of protein (bovine serum albumin, BSA) fouling. Kelly and Zydney thought that the initial fouling for BSA was caused by the convective mass transfer to deposit on the membrane surface. Then BSA can be attached to the existing BSA through chemical attachment via the intermolecular disulfide bonds. It is worth pointing out that due to the deposition of BSA onto the membrane surface, some membrane pores could be blocked, but Kelly and Zydney thought there still existed some small finite flows through these "blocked pores" (Fig. 14).

Consequently, the filtrate flow rate can be written as

$$Q = Q_{\text{open}} + Q_{\text{blocked}} \tag{120}$$

where Q_{open} is the filtration flow rate through open area and $Q_{blocked}$ is the filtration flow rate through blocked area. Apparently, the filtrate flow rate can be predicted if Q_{open} and Q_{blocked} can work out, respectively.

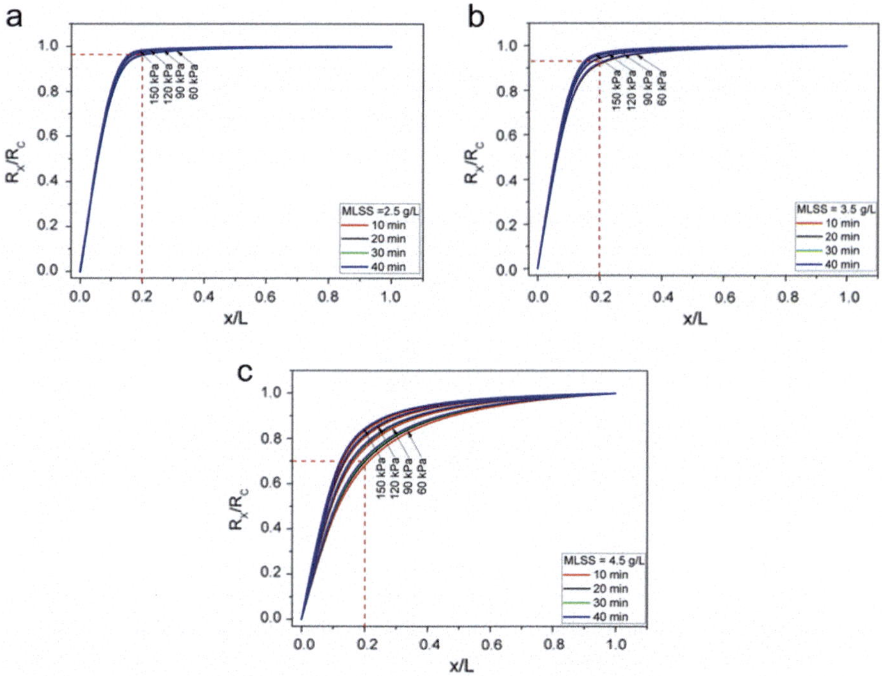

Fig. 12 The cake resistance varies with the relative cake layer thickness (*x*/*l*)

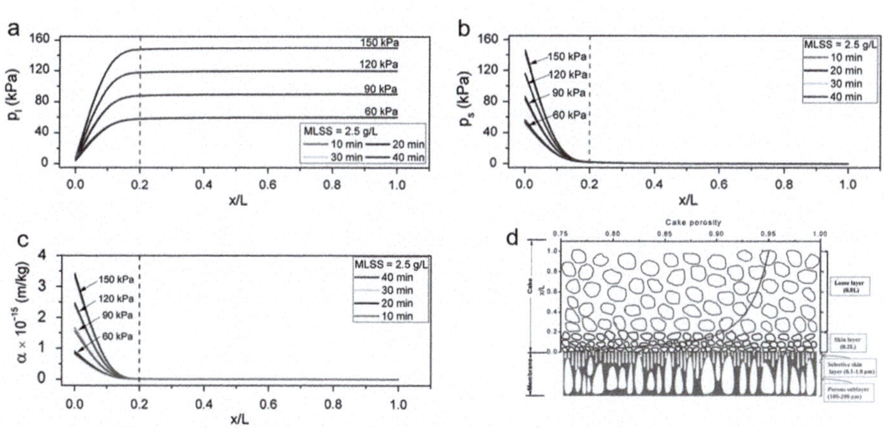

Fig. 13 Spatiotemporal variation of the cake properties

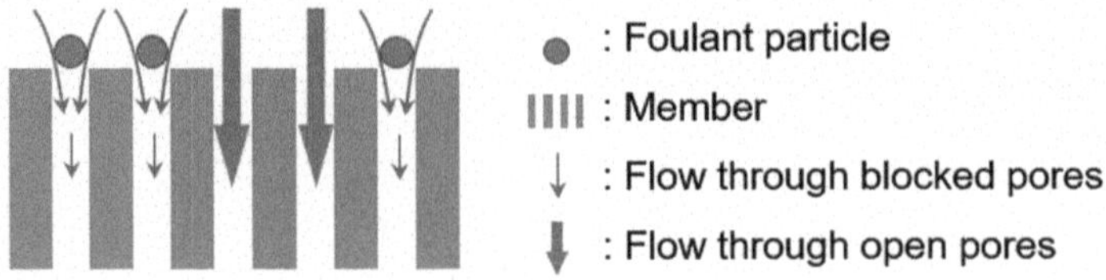

Fig. 14 Diagram of Kelly-Zydney model. A finite flow exists at the blocked pores

Derivation process for Q_{open}

Q_{open} is the volumetric flow rate through the open pores, which is directly related to the fraction of the pores that remain open at any time t.

$$Q_{\text{open}} = \frac{N_0'}{N_0} Q_0 \tag{121}$$

where N_0' is the quantity of the membrane pores remain open, N_0 is the initial quantity of the membrane pores, and Q_0 is the initial volumetric flow rate.

Because of the aggregate deposition and chemical attachment, the membrane pores are blocked. The number of open pores can be obtained from mass balance. Assuming that the variation rate of the number of open pores is proportional to the opposite number of Q_{open},

$$\frac{dN_0'}{dt} = -\alpha' C_b Q_{\text{open}} - \beta^0 C_b Q_{\text{open}} N_{\text{blocked}} \tag{122}$$

where α' is the pore-blocking parameter (it is related to the aggregate deposition), β^0 is proportional to the convective transport of native protein toward to the membrane surface, C_b is the bulk protein concentration, Q_{open} is the volumetric flow rate through open area, and N_{blocked} is the quantity of blocked pore ($N_0 - N_0'$). The number of open pores can be obtained after integrating Eq. (122) from $t = 0$ $N = N_0$ to $t = t$ $N = N_0'$.

$$\frac{N_0'}{N_0} = \frac{\frac{\beta^0 N_0}{\alpha'} + 1}{\frac{\beta^0 N_0}{\alpha'} + \exp\left[C_b \frac{Q_0}{N_0}\left(\alpha' + \beta^0 N_0\right)t\right]} \tag{123}$$

Detailed process for solving Eq. (122).

N_{blocked} is the quantity of blocked pore which is equal to $N_0 - N_0'$, and thus Eq. (122) can be rewritten as

$$\frac{dN_0'}{dt} = -\alpha' C_b \frac{N_0'}{N_0} Q_0 - \beta^0 C_b \frac{N_0'}{N_0} Q_0 \left(N_0 - N_0'\right) \tag{122a}$$

Rearranging Eq. (122a):

$$\frac{dN_0'}{dt} = -C_b \frac{N_0'}{N_0} Q_0 \left[\alpha' + \beta^0 (N_0 - N_0')\right] \tag{122b}$$

$$\frac{dN_0'}{N_0'(\alpha' + \beta^0 N_0 - \beta^0 N_0')} = -C_b \frac{Q_0}{N_0} dt \tag{122c}$$

$$\frac{1}{\alpha' + \beta^0 N_0}\left(\frac{1}{N_0'} + \frac{\beta^0}{\alpha' + \beta^0 N_0 - \beta^0 N_0'}\right) dN_0' = -C_b \frac{Q_0}{N_0} dt \tag{122d}$$

$$\left(\frac{1}{N_0'} + \frac{1}{\alpha'/\beta^0 + N_0 - N_0'}\right) dN_0' = -C_b \frac{Q_0}{N_0}(\alpha' + \beta^0 N_0) dt \tag{122e}$$

Integrating Eq. (122e) from $t = 0\ N\ = N_0$ to $t\ = t\ N_0 - N_0'$:

$$\ln\left(\frac{N_0'}{\alpha' + \beta^0 N_0 - \beta^0 N_0'} \frac{\alpha'}{N_0}\right) = -C_b \frac{Q_0}{N_0}(\alpha' + \beta^0 N_0) t \tag{122f}$$

$$\frac{N_0'}{N_0} = \frac{\frac{\beta^0 N_0}{\alpha'} + 1}{\frac{\beta^0 N_0}{\alpha'} + \exp\left[C_b \frac{Q_0}{N_0}(\alpha' + \beta^0 N_0) t\right]} \tag{123}$$

Thus, substituting Eq. (123) into Eq. (121) and the Q_{open} can be solved out:

$$Q_{\text{open}} = Q_0 \frac{\frac{\beta^0 N_0}{\alpha'} + 1}{\frac{\beta^0 N_0}{\alpha'} + \exp\left[\frac{Q_0}{N_0} C_b \alpha' \left(1 + \frac{\beta^0 N_0}{\alpha'}\right) t\right]} \tag{124}$$

Derivation process for Q_{blocked}

The volumetric flow rate through the blocked pores is written as

$$Q_{\text{blocked}} = N_{\text{blocked}} \frac{\Delta P}{R_c} = Q_{\text{ss}} \frac{N_{\text{blocked}}}{N_0} \tag{125}$$

where Q_{ss} is the quasi-steady flux through the completely fouled membrane.

While the quantity of blocked pores is easily obtained:

$$\frac{N_{\text{blocked}}}{N_0} = 1 - \frac{N_0'}{N_0} \tag{126}$$

Here, substituting Eq. (123) into Eq. (126) to achieve the detailed expression about N_{blocked}.

$$\frac{N_{\text{blocked}}}{N_0} = \frac{\exp\left[\frac{Q_0}{N_0} C_b \alpha' \left(1 + \frac{\beta^0 N_0}{\alpha'}\right) t\right] - 1}{\frac{\beta^0 N_0}{\alpha'} + \exp\left[\frac{Q_0}{N_0} C_b \alpha' \left(1 + \frac{\beta^0 N_0}{\alpha'}\right) t\right]} \tag{127}$$

Thus, the volumetric flow rate through the blocked pores is

$$Q_{\text{blocked}} = Q_{\text{ss}} \frac{\exp\left[\frac{Q_0}{N_0} C_b \alpha' \left(1 + \frac{\beta^0 N_0}{\alpha'}\right) t\right] - 1}{\frac{\beta^0 N_0}{\alpha'} + \exp\left[\frac{Q_0}{N_0} C_b \alpha' \left(1 + \frac{\beta^0 N_0}{\alpha'}\right) t\right]} \tag{128}$$

The expression of Kelly-Zydney model

After combining Eqs. (120), (124), and (128), the Kelly-Zydney model can be obtained:

$$\frac{Q}{Q_0} = \frac{\frac{\beta^0 N_0}{\alpha'} + 1 - \frac{Q_{\text{ss}}}{Q_0}\left[1 - \exp\left[\frac{Q_0}{N_0} C_b \alpha' \left(1 + \frac{\beta^0 N_0}{\alpha'}\right) t\right]\right]}{\frac{\beta^0 N_0}{\alpha'} + \exp\left[\frac{Q_0}{N_0} C_b \alpha' \left(1 + \frac{\beta^0 N_0}{\alpha'}\right) t\right]} \tag{129}$$

2.2.6 Modified Pore-Blocking Law (Polyakov Model)

Uniform pore size is assumed in the classical pore-blocking mechanism. However, actual membrane pores have different radii and bring about different permeability and sieving ability. Polyakov [53] proposed a model to describe the permeate flux varies with filtration time, considering the effect of the membrane pore size distribution on the permeability and sieving ability (Fig. 15a).

Knowledge of the probability density function in the fouling process is required before the model derivation. The probability density function is to describe the output of random variables, and it reflects the probability of things that are described. The integral of the probability density function between the total integral interval is equal to 1. For the membrane fouling process, two aspects of the probability density function should be noticed (Fig. 15b):

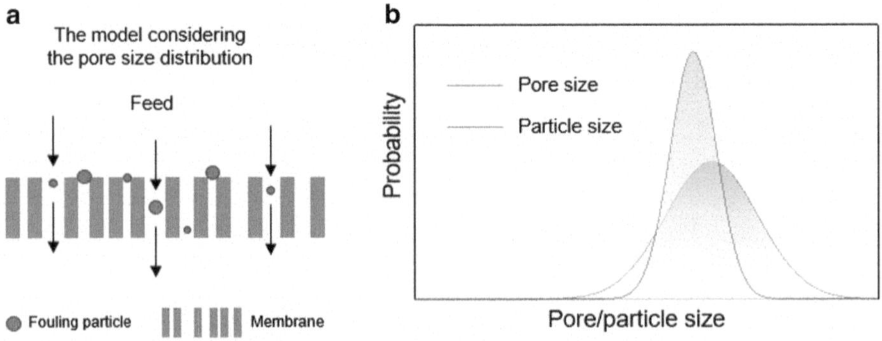

Fig. 15 **a** Membrane with the pore size distribution and its selective performance and **b** the probability density function for the membrane fouling process

$$\int_{r_p^{\min}}^{r_p\max} f_n(r_p, t)dr_p = 1 \qquad (130)$$

$$\int_{r_s^{\min}}^{r_s^{\max}} f_s(r_s)dr_s = 1 \qquad (131)$$

where f_n and f_s are, respectively, the pore and solute probability density functions, r_p and r_s are the pore radius and solute radius, respectively. The superscripts min and max represent the minimum or maximum of the corresponding radius.

According to the classical complete pore-blocking law and Hagen-Poiseuille law, the permeate flux can be written as

$$J = \frac{\pi r_p^4 \Delta P}{8\mu L} N'(t) \qquad (132)$$

Considering the membrane pore size distribution as a major factor in the fouling process, Eq. (132) can be rewritten as

$$J = \frac{\pi \Delta P}{8\mu L} \int_{r_p^{\min}}^{r_p^{\max}} n(r_p, t) r_p^4 \, dr_p \qquad (133)$$

Here, $n(r_p, t)$ is the pore concentration density which can be presented as

$$n(r_p, t) = N'(t) f_n(r_p, t) \qquad (134)$$

If considering $n(r_p, t)$ as a function of r_p, the filtration flux varies with filtration time under the effect of the pore size distribution which can be obtained by substituting $n(r_p, t)$ into Eq. (133). Santos and co-workers obtained the pore plugging kinetics and Polyakov developed the pore plugging kinetics satisfying the corresponding situation:

$$\frac{dn(r_p, t)}{dt} = -\frac{J n(r_p, t) r_p^4}{\int_{r_p^{\min}}^{r_p\max} n(r_p, t) r_p^4 dr_p \int_{r_p^{\min}}^{r_p\max} C f_s(r_s) dr_s} \qquad (135)$$

where C is the total number of solutes per unit volume of solution. Substituting Eq. (133) into Eq. (135) to eliminate the permeate flux (J)

$$\frac{dn(r_p,t)}{dt} = -\frac{\pi \Delta P}{8\mu L} n(r_p,t) r_p^4 \int_{r_p^{\min}}^{r_p \max f} Cf_s(r_s) dr_s \quad (136)$$

Up to now, $n(r_p, t)$ as a function of r_p can be obtained by integrating Eq. (136):

$$n(r_p,t) = n(r_p,0) \exp\left[-\Lambda r_p^4 \int_{r_p^{\min}}^{r_p \max f} Cf_s(r_s) dr_s\right] \quad (137)$$

where $\Lambda = \pi \Delta P / 8\mu L$ and $n(r_p, 0)$ is the pore concentration density at initial time $t = 0$. So far, $n(r_p, t)$ is obtained and the filtration flux varies with filtration time under the effect of the pore size distribution which can be achieved as

$$J = \Lambda \int_{r_p^{\min}}^{r_p^{\max}} n(r_p,0) \exp\left[-\Lambda r_p^4 \int_{r_p^{\min}}^{r_p^{\max}} Cf_s(r_s) \mathrm{d}r_s\right] r_p^4 \mathrm{d}r_p \quad (138)$$

2.2.7 Wang Model for Unsteady/Steady-State Cross-Flow Hollow-Fiber Module

The membrane fouling usually becomes more severe in the dead-end filtration process. However, in the cross-flow filtration process, the membrane fouling could reach equilibrium due to the shear stress, especially for the hollow-fiber module. Thus, it is urgent to figure out the fouling behavior for the cross-flow operating mode in the hollow-fiber module. Wang and co-workers have done this work by dividing the process into two stages: unsteady state [54] and steady state [55], and the corresponding models were proposed according to the mass balance.

For the unsteady-state case, the hollow-fiber module is shown in Fig. 16. Assuming that the average thickness inside the hollow-fiber membrane surface is related to the mass deposit on the membrane surface and the mass resuspends from the foulant layer on the membrane surface by shear force:

$$\frac{d\delta_d(t)}{dt} = BC_b \frac{J}{J_0} - \varsigma^{n''}(t)\delta_d(t) \quad (139)$$

where $\delta_d(t)$ is the average foulant layer thickness on the inside surface of the hollow fiber, m; B is the parameter; C_b is the bulk concentration, kg·m^{-3}; $\varsigma(t)$ is the average surface wall shear rate in the hollow fiber, s^{-1}; and n'' is an experiential constant.

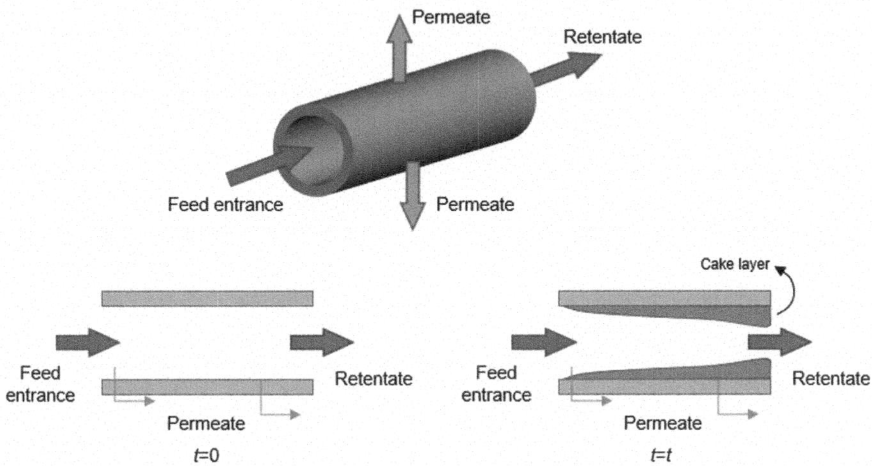

Fig. 16 Diagram of unsteady-state case for hollow-fiber module

It is seen that the first term on the right side of Eq. (139) presents the foulant deposit onto the inside surface of the hollow fiber, and the second term presents the foulant resuspend from the foulant layer by shear force.

The average shear rate $\zeta(t)$ is established and written as

$$\zeta(t) = \frac{2\mu}{\rho R_{\text{in}}^2}\text{Re} = \frac{2\mu}{\rho R_{\text{in}}^2}\frac{2R_{\text{in}}u(t)\rho}{\mu} = \frac{4}{R_{\text{in}}}u(t) \tag{140}$$

where R_{in} is the internal radius of the hollow fiber, m and $u(t)$ is the average flow velocity of the feed solution/suspension in the hollow fiber, m·s^{-1}.

Here, the foulant layer thickness was considered a function of filtration time. Next, the connection between Darcy's law and foulant layer thickness was established. Fortunately, the specific cake resistance of height can be helpful, and Darcy's law can be rewritten as

$$J = \frac{\Delta P}{\mu(R_m + R_d(t))} = \frac{\Delta P}{\mu(R_m + r_d\delta_d(t))} \tag{141}$$

where $R_d(t)$ is the average resistance of foulant deposit on the inside surface of the hollow fiber, m^{-1}. R_d is the specific cake resistance of unit thickness of foulant layer, m^{-2}.

Boundary should be given to solve this first-order differential equation:

$$t = 0,\ \delta_d(t) = 0,\ J = J_0 \tag{142}$$

Thus, the parameter B can be solved when $t = 0$:

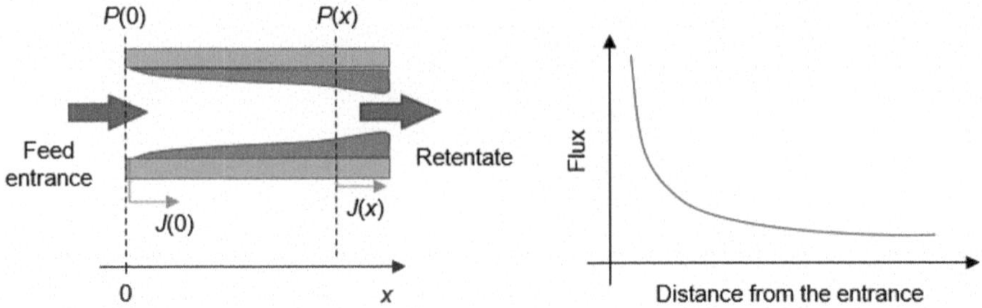

Fig. 17 Diagram of steady-state case for hollow-fiber module and the flux variation with distance

$$B = \frac{1}{C_b} \frac{d\delta_d(t)}{dt}\bigg|_{t=0} \tag{143}$$

After differential process for Eq. (141) and the result is substituted into Eq. (143), B can be rewritten as

$$B = -\frac{1}{C_b} \frac{1}{r_d} \frac{\Delta P}{\mu} \frac{1}{J_0^2} \frac{dJ}{dt}\bigg|_{t=0} \tag{144}$$

In such a way, the unsteady-state case of the cross-flow operating mode for the hollow-fiber module was solved entirely. As the filtration goes on, this unsteady state must disappear, and a steady state must come due to the shear force causing the equilibrium of deposit and resuspend of the surface foulant layer. In this situation, the permeate flux is no longer related to the time, but the pressure drop along the hollow fiber (Fig. 17).

The permeate flux along the fiber at steady state can be written as

$$J = \frac{\Delta P}{\mu(R_m + R_d(x))} = \frac{P(x) - P_0}{\mu(R_m + r_d\delta_d(x))} \tag{145}$$

where $P(x)$ is the pressure along the fiber at distance x from the entrance of the hollow-fiber module, Pa and P_0 is the atmospheric pressure, Pa.

Meanwhile, the permeate flow can be expressed by the Hagen-Poiseuille equation:

$$-\frac{dP(x)}{dx} = \frac{8\mu q(x)}{N\pi R_{in}^4} \tag{146}$$

where $q(x)$ is the volume flow rate along the fiber, $m^3 \cdot s^{-1}$.

Apparently, the expression of the resistance $R_d(x)$ is similar with unsteady state:

$$R_d(x) = r_d\delta_d(x) \tag{147}$$

$$\frac{d\delta_d(x)}{dx} = A'C_b\frac{J(x)}{J_0} - K'\zeta^n(x)\delta_d(x) \tag{148}$$

$$\zeta(x) = \frac{2\mu}{\rho R_{in}^2}\mathrm{Re} = \frac{2\mu}{\rho R_{in}^2}\frac{2R_{in}u(x)\rho}{\mu} = \frac{4}{R_{in}}u(x) \tag{149}$$

where $A^{'}$ and $K^{'}$ are coefficients for the steady-state case.

Now, giving the boundary of these equations:

$$x = 0,\ \delta_d(0) = 0 \tag{150}$$

The coefficients $A^{'}$ can be solved as

$$A' = \frac{1}{C_b}\frac{d\delta_d(x)}{dt}\bigg|_{x=0} \tag{151}$$

Due to the fiber length that is much bigger than R_{in}, it is easy to understand that the variation of foulant layer thickness at the end of the fiber is close to zero:

$$\frac{d\delta_d(x)}{dx}\bigg|_{\mathrm{end}} = 0 \tag{152}$$

Thus, $K^{'}$ can be solved out by substituting Eqs. (151) and (152) into Eq. (148):

$$K' = \frac{AC_bJ(x)}{J_0}\zeta^{-n}(x)\delta_d^{-1}(x)\bigg|_{\mathrm{end}} = \frac{AC_bR_m}{R_m + r_d\delta_d(x)}\zeta^{-n}(x)\delta_d^{-1}(x)\bigg|_{\mathrm{end}} \tag{153}$$

So far, the unsteady/steady-state cross-flow mode model for the hollow-fiber module has been successfully obtained. Moreover, the model for the whole fouling process was successfully established in unsteady/steady-state cross-flow mode.

2.2.8 Models for Plated Device [56, 57]

Plated device is a commonly used membrane device. The membrane channel's upper and lower surfaces are working surfaces (Fig. 18). To get a deep understanding of the ultrafiltration behaviors in the plate and frame module, Wang and co-workers carried out numerous experiments. Finally, at the end of the twentieth century, with the aid of fluid mechanics, Wang et al. proposed mathematical models to describe the plated ultrafiltration process at an unsteady and steady state.

Pulsatory movement of the feed solution/suspension

According to the laws of conservation of matter and momentum, a series of equations could be obtained to describe the fluid flow of two-dimensional incompressible Newtonian fluid under an unsteady state:

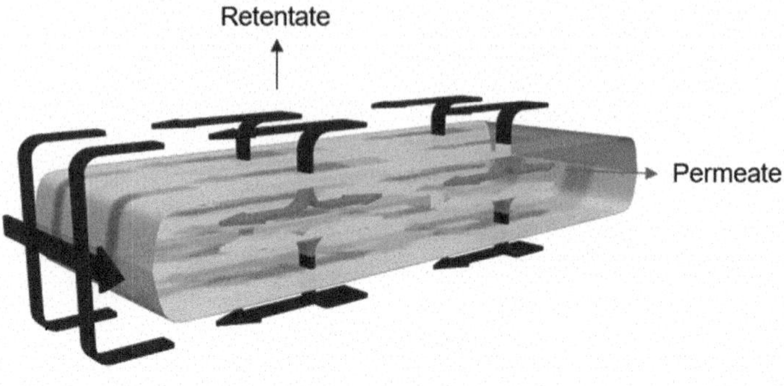

Fig. 18 Sketch map of plated ultrafiltration device

$$\frac{\partial u}{\partial x} + \frac{\partial v}{\partial y} = 0 \tag{154}$$

$$\frac{\partial v}{\partial t} + u\frac{\partial v}{\partial x} + v\frac{\partial v}{\partial y} = -\frac{1}{\rho}\frac{\partial p}{\partial x} + \frac{\mu}{\rho}\left(\frac{\partial^2 v}{\partial x^2} + \frac{\partial^2 v}{\partial y^2}\right) \tag{155}$$

$$\frac{\partial u}{\partial t} + u\frac{\partial u}{\partial x} + v\frac{\partial v}{\partial y} = -\frac{1}{\rho}\frac{\partial p}{\partial x} + \frac{\mu}{\rho}\left(\frac{\partial^2 u}{\partial x^2} + \frac{\partial^2 u}{\partial y^2}\right) \tag{156}$$

$$u|_{y=\pm hf} = 0 \tag{157}$$

$$v|_{y=\pm hf} = v_w \tag{158}$$

$$\left.\frac{\partial v}{\partial y}\right|_{y=0} = 0 \tag{159}$$

$$Q = \int_{-h}^{h} u|_{x=0} dy = \overline{Q}_0(1 + A_m \cos \omega t) \tag{160}$$

$$\overline{Q}_0 = \frac{1}{2}(Q_{\max} + Q_{\min}) \tag{161}$$

$$\mathrm{Am} = (Q_{\max} - Q_0)/Q_0 \tag{162}$$

where u and v are the flow velocity along the x-axis and y-axis, respectively; ρ is the fluid density; μ is the fluid viscosity; h_f is the half height of membrane channel; v_w is the flow velocity near the membrane surface; Q, $\overline{Q}_0$, $Q_{\max}$, and $Q_{\min}$ are the feed volume flow rate, the average feed volume flow rate, the maximum feed volume flow rate, and the minimum feed volume flow rate, respectively.

In the case of pulsatory movement of feed, the flow velocity and pressure can be expressed as:

$$u = u_0(y) + u_1(y)e^{-\mathrm{it}} \tag{163}$$

$$-\frac{\partial p^{(1)}}{\partial x} = a_0(x) + a_1(x)e^{-it} \tag{164}$$

where $u_0(y)$ and $u_1(y)$ are the dimensionless velocity component under steady state and unsteady state, respectively; $a_0(x)$ and $a_1(x)$ are the dimensionless pressure component under steady state and unsteady state.

Solving the above equations by the series expansion method, the maximum shear force during a pulsation period can be obtained as follows:

$$\zeta_{\max} = \sqrt{a_{1c}^2 + [a_{1s} + a^2 A_m]^2} \tag{165}$$

$$a_{1c} = \frac{F_1 B_2}{B_1^2 + B_2^2},\ a_{1s} = \frac{F_1 B_1}{B_1^2 + B_2^2}$$

$$F_1 = 2A_m\beta^2,\ \beta = h\sqrt{\frac{\omega\rho}{2\mu}},\ \alpha = h\sqrt{\frac{\omega\rho}{\mu}}$$

$$B_1 = 1 - \frac{\sin 2\beta + sh2\beta}{2\beta(\cos 2\beta + ch2\beta)},\ B_2 = \frac{sh2\beta - \sin 2\beta}{2\beta(\cos 2\beta + ch2\beta)}$$

In plated ultrafiltration device, the distribution of shear force determines the fouling situation at each place of the membrane module. Based on the maximum shear force above, the filtration flux for plated ultrafiltration device can be obtained as follows:

$$J(x) = \frac{\Delta p}{\mu\left(R_0 + R_c + R_p'\right)} \tag{166}$$

$$u(x) = u(0) - \frac{2}{h}\int_0^x J(x)dx \tag{167}$$

$$\frac{d\delta(x)}{dx} = A''[J(x) - K''\zeta_{\max} \tag{168}$$

$$\delta(0) = 0 \tag{169}$$

where R_1 is the membrane resistance, R_c is the resistance of substance deposited on the membrane surface (cake dominated), $R_p^{'}$ is the concentration polarization resistance, $J(x)$ is the filtration flux along the x-axis, $\delta(x)$ is the thickness of deposited substance, and A'' and K'' are the model parameters.

Steady-state movement of the feed solution/suspension

For steady-state process, the only one difference is the expression of shear force:

$$\zeta = \frac{\mu}{2h^2\rho} C_f \mathrm{Re}^2 \tag{170}$$

where C_f is the effective drag coefficient and *Re* is the Reynolds number.

It should be pointed out that the length of the membrane channel is a non-ignorable factor and there must exist pressure gradient along the membrane channel. Here, the pressure distribution along the membrane channel can be expressed as

$$p(x) = p_0 \left\{ 1 - \frac{\rho u^2(0)}{2p_0} \frac{x}{h} \left[\begin{array}{l} C_f \left(1 - 2\frac{J(x)}{u^2(0)} \frac{x}{h} + \frac{4}{3} \left(\frac{J(x)}{u^2(0)} \right)^2 \left(\frac{x}{h} \right)^2 \right) \\ -4\frac{J(x)}{u^2(0)} \left(1 + \frac{J(x)}{u^2(0)} \frac{x}{h} \right) \end{array} \right] \right\} \tag{171}$$

Application of model

Wang et al. conducted experiments in a plate and frame module with a membrane channel length of 0.931 m, a membrane channel height of 0.46 mm, and membrane materials of polyamide-100000. The experimental results exhibited that this model can accurately describe the filtration behaviors under steady state (Fig. 19) and unsteady state (Fig. 20) for suspension, colloidal substance, and emulsion. Meanwhile, pulsatory movement of feed suspension/solution can significantly alleviate the membrane fouling (Fig. 21).

2.2.9 Force Analysis During Membrane Fouling

Force analysis is a significant method to explore the membrane fouling behavior. Hwang gave an excellent conclusion on force acting on fouling particles [58, 59]. The simple force analysis can be gained as shown in Fig. 22. The forces acting on a particle in vertical direction include drag force induced by permeate flow F_p, net gravitational force F_g, and the lift force F_l and shear-induced force F_s in stirred dead-end or cross-flow filtration.

$$F_p = 3\pi \mu d_p J \tag{172}$$

$$F_g = \frac{\pi}{6} d_p^3 (\rho_s - \rho) \tag{173}$$

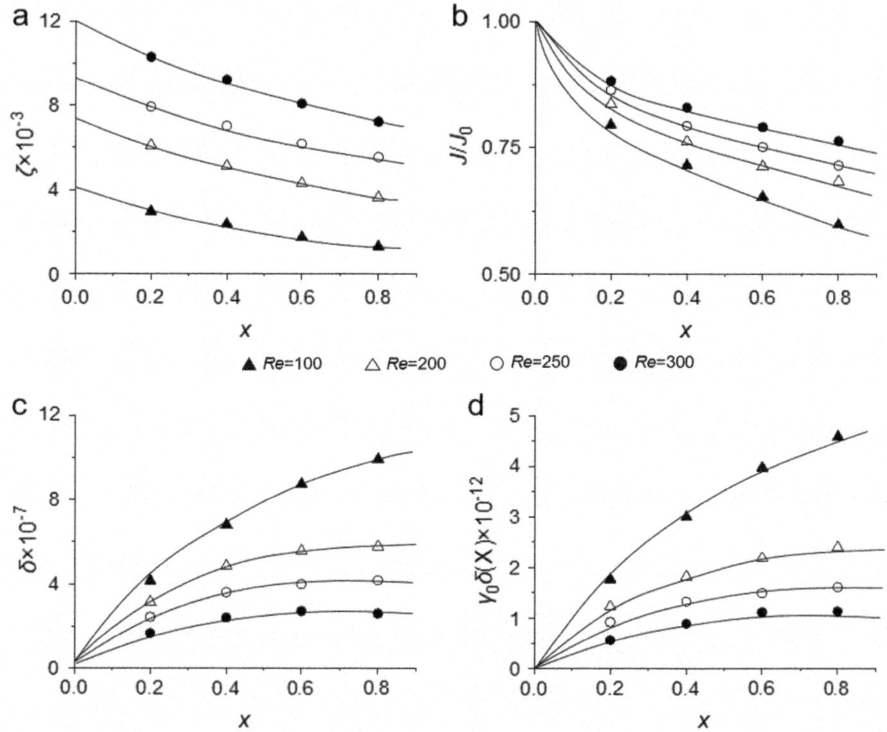

Fig. 19 The effect of Re on the surface shear force, **a** the filtration behavior, **b** deposition foulant thickness, **c** and foulant resistance, **d** along the x

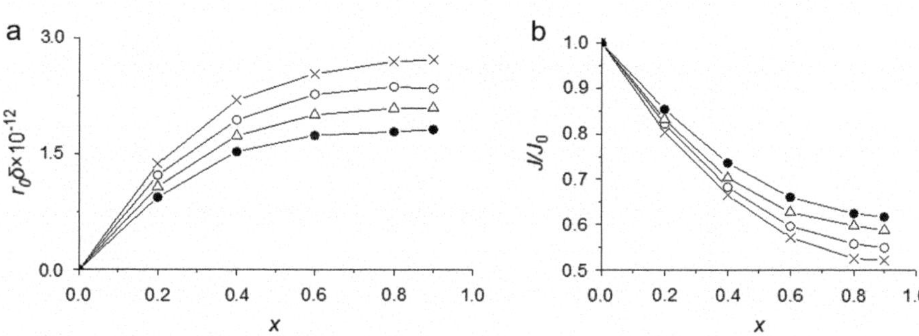

Fig. 20 The effect of $A_m(\omega)$ on the deposition resistance **a** and the filtration behaviors, **b** along the x

where d_p is the fouling particle diameter, F_p is the drag force induced by permeate flow, and F_g is the net gravitational force.

The cross-flow case requires lift force should be analyzed:

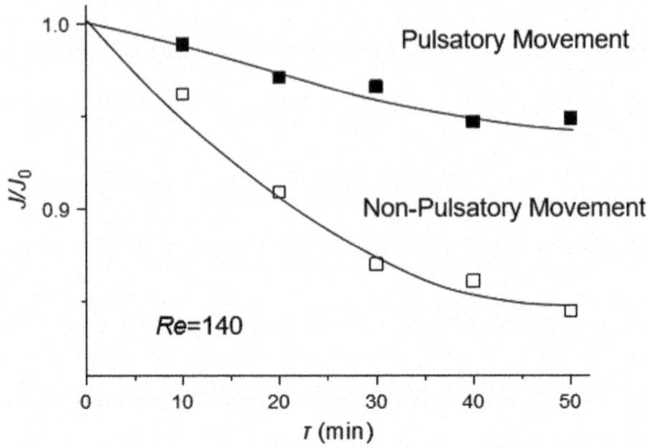

Fig. 21 The effect of pulsatory movement on the membrane filtration behavior

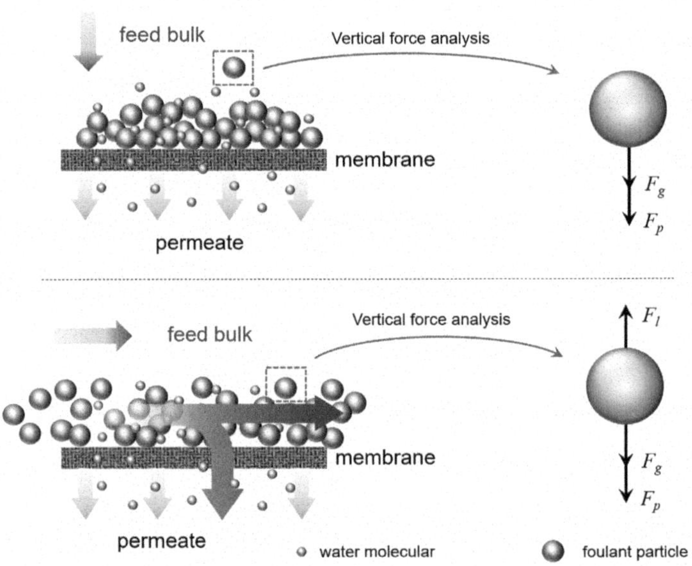

Fig. 22 Force analysis in the membrane fouling process (dead-end and cross-flow mode)

$$F_l = 3\pi \mu d_p u \tag{174}$$

where u is the cross-flow velocity.

However, in micro-scale or even nano-scale, the interaction forces between particle and particle or the membrane also play important roles. Considering the

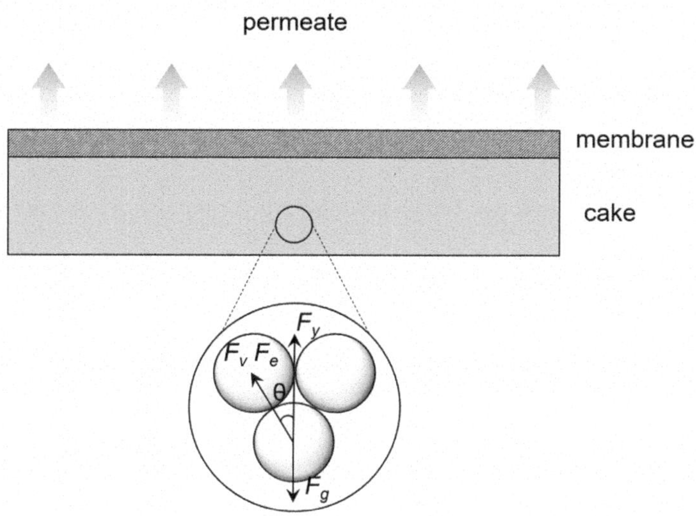

Fig. 23 Schematic representation of force analysis

influence of spatial position, Hwang proposed a force analysis method for steady state which is shown in Fig. 23. To begin with, three assumptions are made:

(1) In unsteady state, particles cannot be resuspended.
(2) The cake layer thickness is constant.
(3) Particles can only form cake but not penetrate into or through the membrane.

Hwang considered the Brownian diffusion force F_b, the Van der Waals attractive force F_v, and double-layer repulsion force F_e between the neighboring particles. Thus, the total force is

$$F = F_y - F_g - K_1(F_v + F_e) \tag{168}$$

where $\vec{F}_y = \vec{F}_p + \vec{F}_b + \vec{F}_s + \vec{F}_l$, $K_1 = 3\cos\theta$.

$$F_y = 3\pi\mu d_p\left(u_p - u_l - u_s - u_b\right)\left[\frac{(R_m + R_c)d_p}{3} + 1.072^2\right]^{0.5} \tag{169}$$

where d_p is the fouling particle diameter; ρ_s is the particle density; u is velocity and their corresponding subscripts *p, l, s*, and *b* mean the permeate direction, the lateral lift direction, the shear-induced velocity direction, and the Brownian diffusion direction, respectively.

$$u_l = \frac{2.6\overline{u}^2 d_p^3 \rho}{16\mu L^2} \tag{170}$$

$$u_d = \frac{\overline{u} d_p^2}{20L^2} \tag{171}$$

$$u_b = \frac{kT}{3\pi \mu L d_p} \tag{172}$$

The van der Waals attractive force and double-layer repulsion force are

$$F_v = -\frac{A d_p^6}{6R^2(R+2d_p)^2(R+d_p)^3} \tag{173}$$

$$F_e = -\left[\frac{\pi d_p^2 \varepsilon_e \varepsilon_0 \zeta^2 \exp(-R/x_a)}{R+d_P}\right]\left(\frac{1}{R+d_P}+\frac{1}{x_a}\right) \tag{174}$$

For the steady state, the net force is zero (i.e., $F = 0$). Thus, permeate velocity can be solved according to Eqs. (168–174)

$$\begin{aligned} u_{ps} = & \frac{2.6\bar{u}^2 d_p^3 \rho}{16\mu L^2} + \frac{\bar{u} d_p^2}{20L^2} + \frac{kT}{3\pi\mu L d_p} + \frac{d_p^2}{18\psi\mu}(\rho_P - \rho) g \\ & - \frac{0.045 A d_p^6}{\psi \mu R^2 (R+2d_p)^2 (R+d_p)^3} \\ & + \frac{0.816 d_p^2 \varepsilon_e \varepsilon_0 \zeta^2 \exp(-R/x_a)}{\psi \mu (R+d_P)}\left(\frac{1}{R+d_P}+\frac{1}{x_a}\right) \end{aligned} \tag{175}$$

where

$$\psi = \left[\frac{(R_m + R_c) d_p}{3} + 1.072^2\right]^{0.5} \tag{176}$$

3 Combined Model

3.1 Applications of Resistance-in-Series Theory

Resistance-in-series theory has been widely used in the field of fluid mechanics, as well as in membrane technology.

In 2002, Carrere and co-workers built a mathematical model based on the resistance-in-series model to describe the membrane performance during the microfiltration of lactic acid fermentation broths [60]. The total resistance was divided into four parts, the membrane resistance (R_m), the cake resistance (R_c), the concentration polarization resistance (Rp), and the adsorption resistance (R_a).

$$J = \frac{p}{\mu\left(R_m + R_c + R_a + R_p\right)} \tag{177}$$

The concentration polarization resistance and adsorption resistance dominated in all cases. Their resistance value could be obtained in the following formula:

$$R_a + R_p = \left(R_{\text{ass}} + R_{\text{pss}}\right) \cdot \left(1 - \exp\left(-b''t\right)\right) \tag{178}$$

where b'' is the constant. R_{ass} and R_{pss} are the adsorption resistance and concentration polarization resistance steady value, respectively.

For the case of the cake compressibility, Carrere used compressibility index n to corrected cake resistance.

$$R_c = \frac{m}{A}\alpha p^n \tag{179}$$

where m/A refers to the bacteria mass in the cake per unit and α is the cake resistance coefficient.

To calculate the cake resistance, following equations should be used:

$$\frac{dm}{dt} = \left(JC - \frac{D}{\delta}(C_w - C)\right)A \tag{180}$$

$$\frac{D}{\delta}C_m = J_{ss}C \tag{181}$$

$$C = \frac{C_0 V_0}{V_0 - A\int_0^t J dt} \tag{182}$$

where C is the bacteria concentration, C_m is the bacteria concentration on the membrane surface, D is the shear-induced diffusion coefficient, and δ is the cake thickness. Owing to the rejection of bacteria, C varies with filtration time. Here, Eq. (181) represents the relationship between C and filtration time. C_0 is the initial bacteria concentration and V_0 is the initial retentate volume.

3.2 *Ho-Zydney Model*

In the actual filtration process, feed solutions/suspensions always have broad particle size distributions, especially for proteins. Adsorption, interaction, and aggregation tend to become more significant because of the intermolecular disulfide bonds [61, 62]. In this case, any single fouling mechanism cannot be satisfied. Meanwhile, numerous types of research provided evidence to demonstrate the transition of the fouling mechanism from pore blocking to cake filtration. It is easy to get this result by assessing the change of parameter n in $\frac{d^2t}{dv^2} = k\left(\frac{dt}{dv}\right)^n$. However, there was a lack of the description of this transition phenomenon and the whole fouling behavior by using one mathematical model till Ho and Zydney proposed a model in 2000 [63].

Based on the concept of Kelly-Zydney model, the new Ho-Zydney model was also established on the small finite flow through the blocked pores. The total filtrate flow rate can also be divided into Q_{open} and Q_{blocked}. Kelly and Zydney only considered the effect of variation of the available membrane pores on the filtrate flow rate while not mentioning the cake layer resistance in the Kelly-Zydney model. Here, Ho and Zydney not only considered the number of open pores but also the resistance of proteins deposited over a particular region (*Rp*). Thus, some expressions can be written as

$$Q = Q_{\mathrm{open}} + Q_{\mathrm{blocked}} \tag{183}$$

$$Q_{\mathrm{open}} = \frac{\Delta P}{\mu R_m} A_{\mathrm{open}} \tag{184}$$

$$Q_{\mathrm{blocked}} = \int\limits_{A_{\mathrm{blocked}}} J_{\mathrm{blocked}} dA = \int\limits_{A_{\mathrm{blocked}}} \frac{\Delta P}{\mu(R_m + R_p)} dA \tag{185}$$

Ho and Zydney firstly give the derivation of Q_{open}. Assuming the rate of pore coverage or blockage is assumed to be proportional to the convective flow rate of protein aggregates to the membrane surface:

$$\frac{dA_{\mathrm{open}}}{dt} = -\alpha' Q_{\mathrm{open}} C_b \tag{186}$$

where C_b is the bulk protein concentration and α' is a pore blockage parameter which is equal to the membrane area blocked per unit mass of protein transfer to the membrane surface.

The pore blockage parameter is related to the fraction amount of total protein present as aggregates (f), the blocked membrane area due to each aggregate (A_{agg}), and the mass of a single aggregate (M_{agg}):

$$\alpha' = \frac{f A_{\mathrm{agg}}}{M_{\mathrm{agg}}} \tag{187}$$

Substituting Eq. (187) into Eq. (186) and integrating from $t = t\ A = A_0$ to $t = t$ $A = A_{\mathrm{open}}$ to obtain the open membrane area as a function of filtration time:

$$A_{\mathrm{open}} = A_0 \exp\left(-\frac{\alpha' \Delta P C_b}{\mu R_m} t\right) \tag{188}$$

Therefore, the filtrate flow rate through the open pores is

$$Q_{\text{open}} = \frac{\Delta P}{\mu R_m} A_0 \exp\left(-\frac{\alpha' \Delta P C_b}{\mu R_m} t\right) = Q_0 \exp\left(-\frac{\alpha' \Delta P C_b}{\mu R_m} t\right) \tag{189}$$

If because the filtrate flows through blocked area Q_{blocked}. The finite flow through the blocked area is ascribed to the resistance proteins deposited on the membrane surface. This resistance is related to the number of proteins on the membrane surface. Thus, it can be written as

$$\frac{dR_p}{dt} = R' \frac{dm_p}{dt} = f' R' J_{\text{blocked}} C_b \tag{190}$$

where $R^{'}$ is the specific cake protein resistance, m_p is the mass of protein deposited onto the membrane surface, and $f^{'}$ is the fractional amount of protein contributed to deposit resistance.

Substituting J_{blocked} of Eq. (185) into Eq. (190) followed by integrating from $t = t_p R_p = R_{p0}$ to $t = t\ R_p = R_p$:

$$R_p = \left(R_m + R_{p0}\right)\sqrt{1 + \frac{2 f' R' \Delta P C_b}{\mu (R_m + R_{p0})^2}\left(t - t_p\right)} - R_m \tag{191}$$

Here, t_p is the time that the region was covered or blocked by proteins firstly.

Now, if we figure out dA_{blocked}, the filtrate flow rate through blocked area Q_{blocked} can be solved out. It is easy to understand that d$A_{\text{blocked}} = -$dA_{open}. According to Eq. (187):

$$dA_{\text{blocked}} = -dA_{\text{open}} = \frac{\alpha' \Delta P C_b}{\mu R_m} A_0 \exp\left(-\frac{\alpha' \Delta P C_b}{\mu R_m} t_p\right) dt_p \tag{192}$$

Thus, substituting Eq. (192) into Eq. (185), Q_{blocked} can be achieved as follows:

$$Q_{\text{blocked}} = Q_0 \int_0^t \frac{\alpha' \Delta P C_b}{\mu (R_m + R_p)} \exp\left(\frac{-\alpha' \Delta P C_b}{\mu R_m} t_p\right) dt_p \tag{193}$$

The final expression of Ho-Zydney model can be obtained by combining Eqs. (183), (189), and (193):

$$Q = Q_0 \left[\exp\left(\frac{-\alpha' \Delta P C_b}{\mu R_m} t\right) + \int_0^t \frac{\alpha' \Delta P C_b}{\mu (R_m + R_p)} \exp\left(\frac{-\alpha' \Delta P C_b}{\mu R_m} t_p\right) dt_p\right] \tag{194}$$

A much simpler analytical solution can be developed for the filtrate flow rate by assuming a uniform resistance of the protein layer over the fouled surface of the membrane. The coefficient multiplying the exponential in the convolution integral in model is a constant and can be pulled outside of the integral to give

$$Q = Q_0\left[\exp\left(\frac{-\alpha' \Delta P C_b}{\mu R_m} t\right) + \frac{R_m}{R_m + R_p} \times \left(1 - \exp\left(\frac{-\alpha' \Delta P C_b}{\mu R_m} t_p\right)\right)\right] \quad (195)$$

3.3 *Bolton Model for Constant Pressure Mode*

Four classical pore-blocking laws have a profound influence on the membrane fouling field. However, the feed solution/suspension contains quantitative foulant particles with the corresponding particle size distribution. A single classical pore-blocking mechanism is unsatisfactory in describing these fouling processes. Bolton and co-workers proposed a novel method to establish numerous combined two-mechanism mathematical models with explicit equations based on the four individual fouling mechanisms [64] (Fig. 24).

Complete pore blocking, standard pore blocking, intermediate pore block, and cake filtration, as four classical pore-blocking mechanisms, can be combined in pairs to obtain a combined two-mechanism model. Then, according to Darcy's law, the flow rate can be written as a function of the filtration resistance and the membrane area.

$$Q = \frac{PA}{R\mu} \quad (196)$$

According to these two terms (the filtration resistance and the membrane area), Bolton divided the four classical pore-blocking mechanisms into two categories:

(1) the complete pore-blocking and intermediate pore-blocking mechanisms can lead to a decrease in the membrane area;
(2) the standard pore-blocking and cake filtration mechanisms can increase the filtration resistance.

Thus, their corresponding expression can be rewritten based on Table 5 as follows:
Complete blocking:

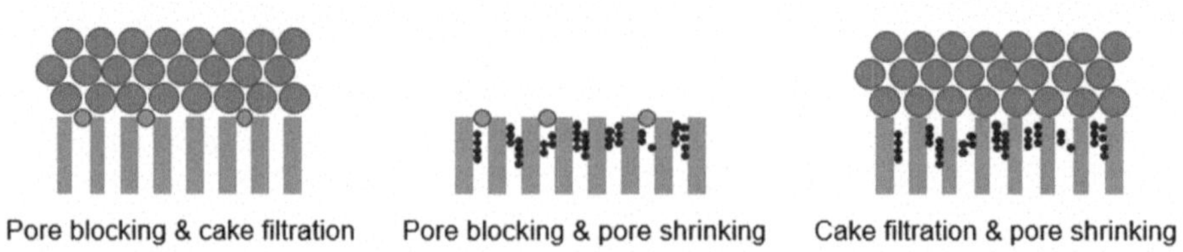

Fig. 24 Diagram of the combined mechanism of pore blocking and cake filtration (**a**), pore blocking and pore shrinking (**b**) and cake filtration and pore shrinking (**c**)

$$\frac{A}{A_0} = 1 - \frac{K_b}{J_0} v \text{or} \frac{A}{A_0} = 1 - K_b t \tag{197}$$

Intermediate blocking:

$$\frac{A}{A_0} = \exp(-K_i v) \text{ or } \frac{A}{A_0} = \exp(-K_i J_0 t) \tag{198}$$

Standard blocking:

$$\frac{R}{R_0} = \left(1 - \frac{K_s v}{2}\right)^{-2} \text{or} \frac{R}{R_0} = \left(1 + \frac{K_s J_0 t}{2}\right)^2 \tag{199}$$

Cake filtration:

$$\frac{R}{R_0} = (1 + K_c J_0 v) \text{ or } \frac{R}{R_0} = \left(1 + 2K_c J_0^2 t\right)^{1/2} \tag{200}$$

where A is the available membrane area at filtration time t, A_0 is the virgin available membrane area, R is the filtration resistance at filtration time t (including the membrane resistance and the foulant resistance), and R_0 is the membrane resistance.

Here, Bolton et al. compared the initial filtration flux and the filtration flux at time t and the equation $J = Q/A_0$:

$$\frac{J}{J_0} = \frac{R_0 A}{R A_0} \tag{201}$$

The initial flow rate and flux:

$$Q_0 = (P A_0)/(R_0 \mu)\, and\, J_0 = \frac{P}{R_0 \mu} \tag{202}$$

The flow rate and flux at time t:

$$Q = \frac{PA}{R\mu}\, and\, J = \frac{PA}{R\mu}\frac{1}{A_0} \tag{203}$$

J/J_0 can be easily obtained by combining Eqs. (202) and (203).

Therefore, the Bolton model with five different models can be obtained by permutating and combining four classical pore-blocking laws in pairs.

(a) **Complete-cake model** (Fig. 24(a))

In this case, both complete pore-blocking and cake filtration happen. To solve this explicit mathematical model, the first-order differential equation about filtrate volume and filtration time for complete pore blocking-cake filtration model can be obtained by substituting Eq. (197) and Eq. (200) into Eq. (201):

$$\frac{J}{J_0} = \frac{R_0 A}{R A_0} = \left(1 + 2K_c J_0^2 t\right)^{-0.5}\left(1 - \frac{K_b}{J_0}v\right) = \frac{dv}{J_0 dt} \tag{204}$$

The integral interval is from $t = 0$ $v = 0$ to $t = t$ $v = v$. After integrating, the filtrate volume as a function of filtration time is

$$v = \frac{J_0}{K_b}\left[1 - \exp\left(-\frac{K_b}{K_c J_0^2}\left((1 + 2K_c J_0^2 t)^{1/2} - 1\right)\right)\right] \tag{205}$$

<u>**Note**</u>:

It should be noted that the choice of *the* A/A_0 term above is a unique solution. Therefore, whether an equation $1 - \frac{K_b}{J_0}v$ or equation $1 - K_b t$ is chosen should not be ambiguous. Furthermore, the cake formation can inhibit the pore-blocking process because the research object is a combined mechanism. That is to say that the available membrane area ratio cannot decrease linearly with filtration time. Thus, the equation $1 - K_b t$ is not available here.

(b) **Intermediate-cake model** (Fig. 24(a))

In this case, the complete pore blocking is replaced by intermediate pore blocking to lead to the available membrane area decrease. The cake formation also contributes to the resistance increase and the flux decay. Similarly, the first-order differential equation about filtrate volume and filtration time for the intermediate pore blocking-cake filtration model can be obtained by substituting Eqs. (198) and (200) into Eq. (201):

$$\frac{J}{J_0} = \frac{R_0 A}{R A_0} = \left(1 + 2K_c J_0^2 t\right)^{-0.5}\exp(-K_i v) = \frac{dv}{J_0 dt} \tag{206}$$

After integrating from $t = 0$ $v = 0$ to $t = t$ $v = v$ and the filtrate volume as a function of filtration time is:

$$v = \frac{1}{K_i}\ln\left[1 + \frac{K_i}{K_c J_0}\left((1 + 2K_c J_0^2 t)^{1/2} - 1\right)\right] \tag{207}$$

(c) **Complete-standard model** (Fig. 24(b))

In this case, complete pore blocking is treated as the factor leading to the available membrane area decrease and standard pore blocking is regarded as the factor leading to the resistance increase. Hence, the first-order differential equation about filtrate volume and filtration time for complete pore blocking-standard pore-blocking model can be obtained by substituting Eqs. (197) and (199) into Eq. (201):

$$\frac{J}{J_0} = \frac{R_0 A}{R A_0} = \left(1 + \frac{K_s J_0 t}{2}\right)^{-2} \left(1 - \frac{K_b}{J_0} v\right) = \frac{dv}{J_0 dt} \tag{208}$$

The integral interval is from $t = 0$ $v = 0$ to $t = t$ $v = v$ and the filtrate volume as a function of filtration time is

$$v = \frac{J_0}{K_b}\left(1 - \exp(\frac{-K_b t}{1 + K_s J_0 t/2})\right) \tag{209}$$

(d) **Intermediate-standard model** (Fig. 21(b))

In this case, intermediate and standard pore blocking are chosen to establish a combined mechanism model. Here, intermediate pore blocking can decrease the available membrane area, and standard pore blocking can lead to an increase in filtration resistance. Similar to the three models described above, the first-order differential equation about filtrate volume and filtration time for intermediate pore blocking-standard pore-blocking model can be obtained by substituting Eqs. (198) and (199) into Eq. (201):

$$\frac{J}{J_0} = \frac{R_0 A}{R A_0} = \left(1 + \frac{K_s J_0 t}{2}\right)^{-2} \exp(-K_i v) = \frac{dv}{J_0 dt} \tag{210}$$

The filtrate volume as a function of filtration time can be obtained by integrating Eq. (210) from $t = 0$ $v = 0$ to $t = t$ $v = v$.

$$v = \frac{1}{K_i} \ln\left(1 + \frac{K_i J_0 t}{1 + K_s J_0 t/2}\right) \tag{211}$$

(e) **Cake-standard model** (Fig. 24(c))

In this case, the available membrane area ratio always keeps one, and only the filtration resistance increases. The cake and internal foulant particles deposited on the pore walls increase the resistance. It should be pointed out that the resistance Eq. (200) described contained not only the cake resistance but also the membrane resistance and Eq. (199). Thus, to express the cake resistance, Eq. (200) should be rewritten as

$$\frac{R_c}{R_0} = \frac{R - R_0}{R_0} = K_c J_0 v \tag{212}$$

Therefore, the first-order differential equation about filtrate volume and filtration time for the cake filtration-standard pore-blocking model can be obtained by substituting Eqs. (199) and (212) into Eq. (201):

$$\frac{J}{J_0} = \left((1 - \frac{K_s v}{2})^{-2} + K_c J_0 v\right)^{-1} = \frac{dv}{J_0 dt} \tag{213}$$

After integrating from $t = 0$ $v = 0$ to $t = t$ $v = v$ and t as a function of filtrate volume v, we can obtain

$$t = \frac{1}{K_s v - 2}\left(\frac{K_s K_c v^3}{2} - K_c v^2 - \frac{2v}{J_0}\right) \tag{214}$$

It is possible to achieve the filtrate volume as a function of filtration time through Eq. (214):

$$v = \frac{2}{K_s}\left[\beta\cos\left(\frac{2}{3}\pi - \frac{1}{3}\text{arc}\ \cos(\alpha)\right) + \frac{1}{3}\right] \tag{215}$$

where $\alpha \left(\frac{8}{27\beta^3} + \frac{4K_s}{3\beta^3 K_c J_0} - \frac{4K_s^2 t}{3\beta^3 K_c}\right)$ and $\beta \left(\frac{4}{9} + \frac{4K_s}{3K_c J_0} + \frac{2K_s^2 t}{3K_c}\right)$ are the parameters that used to simplify the solution.

Thus far, Bolton's five combined models in constant pressure mode have been ultimately introduced. Bolton and co-workers used their model to predict the flux decay for bovine serum album (BSA) in the filtration process. The complete pore blocking-cake filtration model was the best to describe its fouling behavior. Many types of research were carried out based on this research.

3.4 Hou Model

In previous studies [63, 64], the fouling mechanism of BSA on the microfiltration membrane is considered a combined model with complete pore-blocking and cake filtration mechanisms. Meanwhile, numerous models were established to describe the fouling behavior (i.e., the flux decay) of the combined complete pore-blocking and cake filtration mechanism. As discussed above, Bolton analyzed the contribution of the resistance and the available membrane area to the flux decay [64]. However, it is interesting that the available membrane area must decrease to 0 after a specific time for every foulant, which can cause pore blocking in Bolton's research. However, Hou and co-workers thought that as the filtration goes and the filter cake would become denser and denser so that the cake could be used as an umbrella to prevent membrane pores from being blocked by some foulant particles (Fig. 25) [65]. If so, the available membrane area could not decrease to 0. Instead, it could be kept constant after a specific time (i.e., the available membrane area reached steadily). Herein, a parameter K was introduced into the Bolton model to correct the available membrane area ratio [65].

Meanwhile, the fouling mechanism of combined complete pore blocking and cake filtration is inappropriate for describing the whole fouling process. Hou thought that the fouling mechanism was firstly combined (complete pore blocking-cake filtration) and then converted to a single mechanism (cake filtration) (Fig. 25). Therefore, there also existed a transition point for the fouling mechanism change. Thus, the

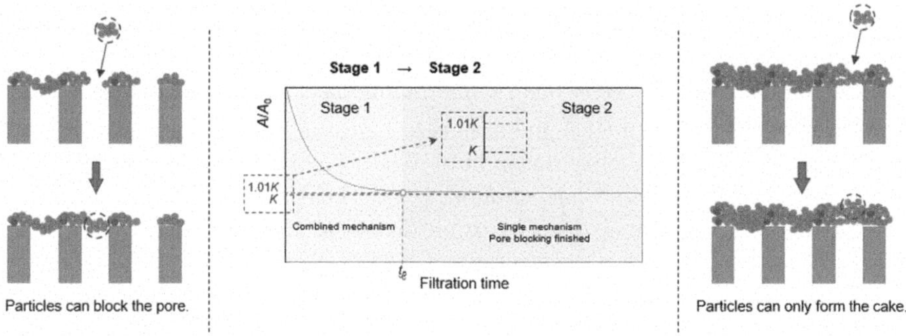

Fig. 25 Diagram of Hou model. Denser cake would be formed to inhibit pore blocking from happening. The variation of A/A_0 can be used to determine transition point of the membrane fouling mechanism

application of boundary layer theory in describing the available membrane ratio was first proposed to determine the transition point (Fig. 25 mediate).

In the fouling process, the available membrane area ratio decreases to a steady value *of* K. Boundary layer theory tells us that if the difference between the available membrane area ratio and K is equal to 1%. In other words, i.e., decreasing to 1.01 K. *The* corresponding time is the transition point when the fouling mechanism is converted from a combined mechanism to a single mechanism.

According to Bolton model, a combined model of complete pore blocking-cake filtration was proposed [65]:

$$J = J_0 \frac{\exp\left(-\frac{K_b}{K_c J_0^2}\left((1+2K_c J_0^2 t)^{1/2}-1\right)\right)}{(1+2K_c J_0^2 t)^{1/2}} \tag{216}$$

$$\frac{A}{A_0} = \exp\left(-\frac{K_b}{K_c J_0^2}\left((1+2K_c J_0^2 t)^{1/2}-1\right)\right) \tag{217}$$

After introducing the parameter K, the available membrane area ratio can be rewritten as follows:

$$\frac{A}{A_0} = (1-K)\exp\left(-\frac{K_b}{K_c J_0^2}\left((1+2K_c J_0^2 t)^{1/2}-1\right)\right) + K \tag{218}$$

Thus, the filtration flux as a function of filtration time can be obtained after combining Eq. (216) with Eq. (218).

$$J = J_0 \frac{(1-K)\exp\left(-\frac{K_b}{K_c J_0^2}\left((1+2K_c J_0^2 t)^{1/2}-1\right)\right) + K}{(1+2K_c J_0^2 t)^{1/2}} \tag{219}$$

It is worth to note that the model prediction accuracy has a remarkable promotion after introducing parameter K.

Based on the boundary layer theory, the fouling mechanism transition point can be obtained when $A/A_0 = 1.01$ K in Eq. (218).

$$t'_p = \frac{1}{2K_c J_0^2}\left[\left(1 - \frac{K_c J_0^2}{K_b}\ln\left(\frac{0.01K}{1-K}\right)\right)^2 - 1\right] \tag{220}$$

3.5 *Yao Model*

The early stage of the filtration process is critical. Therefore, understanding the mechanism of the cake layer formation in the initial and pseudo-steady stages is imperative in MBR. Yao and co-workers focused on the fouling behaviors in the early stage at different carbon-to-nitrogen ratios (C/N) [66]. They proposed a combined mechanism and used a piecewise function to describe the relationship between filtrate volume and filtration time.

Yao et al. thought that in the **initial filtration stage**, sludge flocs are deposited on the membrane surface to form "passageways," followed by the tiny particles of colloid matter entering these "passageways." They formed the colloid layer on the membrane surface in the **earlier pseudo-steady stage**, and then in the **late pseudo-steady stage,** colloid layer was formed entirely on the membrane surface. The schematic diagram of the fouling mechanism is shown in Fig. 26.

The mathematical model can be a piecewise function that divides into three stages. As mentioned above, the **initial filtration stage** can be regarded as a simple cake formation process, and Eq. (30) can be used to describe this process:

$$\frac{t}{v} = \frac{K_{fc}}{2}v + \frac{1}{J_0} \tag{221}$$

where K_{fc} is the flocs cake filtration constant.

The **earlier pseudo-steady stage** can be seen as a combined mechanism of pore shrinking and cake formation. According to Bolton's model of cake-standard mechanism (Eq. (213)), the model can be easily obtained as

$$\frac{t}{V} = \frac{1}{K_h v - 2}\left(\frac{K_h K_{cc} v^2}{2} - K_{cc}v - \frac{2}{J_{h0}}\right) \tag{222}$$

where K_h is the hole constriction constant, K_{cc} is the colloid cake layer filtration constant, and J_{h0} is the initial filtration flux for hole constriction stage.

The **late pseudo-steady stage** is the complete colloid cake formation stage and single cake filtration law can also be used to describe it. Thus, the expression is

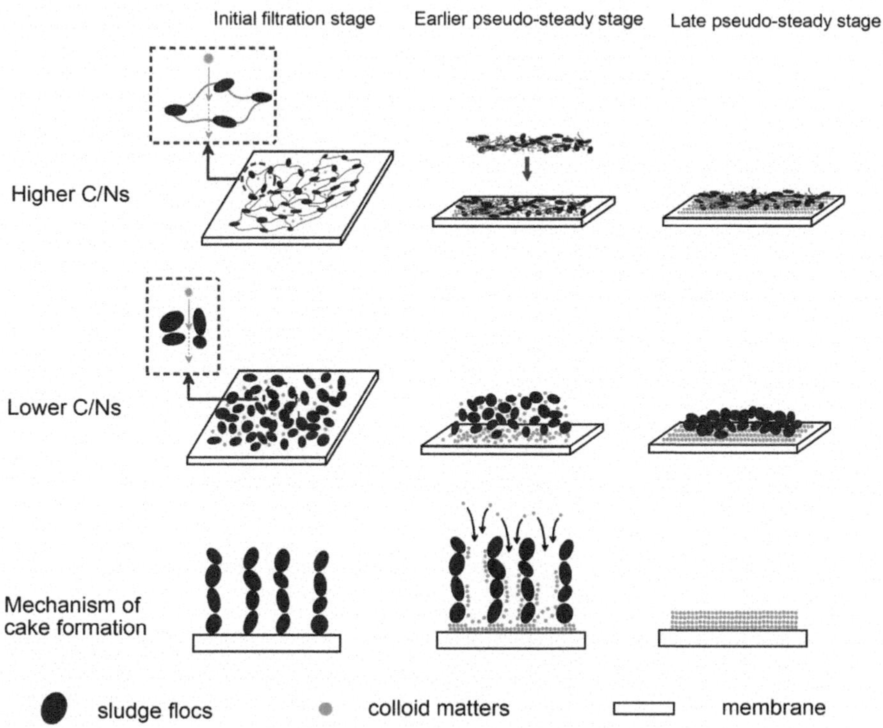

Fig. 26 Mechanism of the membrane fouling in the early stage for MBR

$$\frac{t}{v} = \frac{K_{cc}}{2}v + \frac{1}{J_{c0}} \tag{223}$$

where J_{cO} is the initial filtration flux for colloid cake formation stage.

Herein, the final expression of Yao model can be written as

$$\begin{cases} \frac{t}{V} = \frac{K_{fc}}{2}v + \frac{1}{J_0}t \le t_1 \\ \frac{t}{V} = \frac{1}{K_h v - 2}\left(\frac{K_h K_{cc} v^2}{2} - K_{cc}v - \frac{2}{J_{h0}}\right)t_1 \le t \le t_2 \\ \frac{t}{V} = \frac{K_{cc}}{2}v + \frac{1}{J_{c0}}t_2 \le t \end{cases} \tag{224}$$

where t_1 is the transition time between **initial filtration stage** and **earlier pseudo-steady stage** and t_2 is the transition time between **earlier pseudo-steady stage** and **late pseudo-steady stage**.

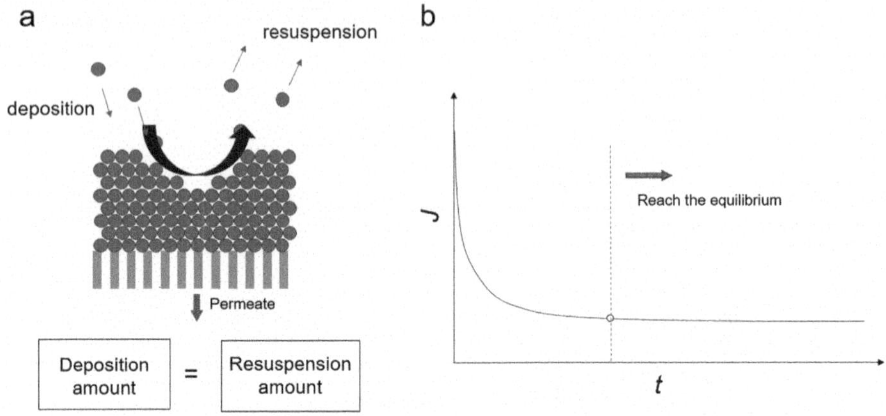

Fig. 27 **a** Equilibrium between foulant deposition amount and resuspension amount in cross-flow filtration case. **b** The flux decreases with filtration time and finally reaches a platform

3.6 *Cake Equilibrium Coefficient Model (Li Model)*

Generally speaking, the shear stress induced by cross flow can alleviate the membrane fouling as it can remove some fouling particles from the membrane surface. However, as the filtration process continues, the cake deposition and resuspension may reach an equilibrium (Fig. 27). Li and co-workers considering this equilibrium of cake amount and a steady state of available membrane area proposed a model to describe the flux decay during the cross-flow microfiltration process [67].

Similarly, the Li model is based on the Bolton concept that the flux variation is related to the resistance and membrane frontal area

$$\frac{J}{J_0} = \frac{R_0 A}{R A_0} \tag{225}$$

Taking into account the combined mechanism of complete pore blocking and cake filtration, the resistance variation and membrane frontal area can be expressed as

$$\frac{R_0}{R} = \left(1 + 2K_c J_0^2 t\right)^{-1/2} \tag{226}$$

$$\frac{A}{A_0} = 1 - \frac{K_b}{J_0} v \tag{227}$$

Owing to the equilibrium between the cake deposition and the cake removal, the resistance cannot permanently rise until reaching a platform. Paradoxically, from the viewpoint of mathematics, Eq. (226) shows that the resistance can increase all along. Herein, Li introduced a cake equilibrium coefficient to correct Eq. (226)

$$\frac{R_0}{R} = (1 - K_d)\left(1 + 2K_c J_0^2 t\right)^{-1/2} + K_d \tag{228}$$

where K_d is the cake equilibrium coefficient.

K_d changes from 0 to 1. Smaller K_d refers to higher cake resistance. Substituting Eqs. (227) and (228) into Eq. (225), we obtain a differential equation:

$$\frac{J}{J_0} = \left(1 - \frac{K_b}{J_0}v\right)\left((1 - K_d)\left(1 + 2K_c J_0^2 t\right)^{-1/2} + K_d\right) = \frac{dv}{J_0 dt} \tag{229}$$

After integrating, the relation between filtration volume and time is

$$v = \frac{J_0}{K_b}\left(1 - \exp\left((K_d - 1)\frac{K_b}{K_c J_0^2}((1 + 2K_c J_0^2 t)^{1/2} - 1) - K_b K_d t\right)\right) \tag{230}$$

Thus, the flux can be written as

$$J = J_0 \exp\left((K_d - 1)\frac{K_b}{K_c J_0^2}((1 + 2K_c J_0^2 t)^{1/2} - 1) - K_b K_d t\right) \left(\frac{1 - K_d}{(1 + 2K_c J_0^2 t)^{1/2}} + K_d\right) \tag{231}$$

In Sect. 3.4, Hou model described that the loss of the membrane frontal area may also reach to a platform because denser cake layer formed to inhibit the pore blocking from happening. Therefore, Li also corrected the membrane frontal area here.

$$\frac{A}{A_0} = (1 - K)\exp\left((K_d - 1)\frac{K_b}{K_c J_0^2}((1 + 2K_c J_0^2 t)^{1/2} - 1) - K_b K_d t\right) + K \tag{232}$$

where K is the steady value of the membrane frontal area ratio.

The expression of Li model can finally be written as

$$J = J_0\left[(1 - K)\exp\left((K_d - 1)\frac{K_b}{K_c J_0^2}((1 + 2K_c J_0^2 t)^{1/2} - 1) - K_b K_d t\right) + K\right] \left(\frac{1 - K_d}{(1 + 2K_c J_0^2 t)^{1/2}} + K_d\right) \tag{233}$$

References

1. Z. Wang, X.P. Du, H. Zhang, X. Wang, P. Song, K. Natsagdorj, B. Khan, R. Khurram, The application of error function for normalized flux prediction in dead-end microfiltration (MF) process. Sep. Sci. Technol. **56**, 117–128 (2021)
2. Z. Wang, J.M. Yao, X. Ye, J.S. Chu, Y.J. Cui, Y.L. Liang, H. Zhang, X.Q. Zhao, Z.H. Li, Influence of operating conditions on ultrafiltration polysaccharide of Agaricus Blazei Murill. J. Beijing Univ. Technol. **34**, 1105–1110 (2008)
3. Z. Wang, J. Yao, C. Zhou, J. Chu, The influence of various operating conditions on the permeation flux during dead-end microfiltration. Desalination **212**, 209–218 (2007)
4. J.S. Chu, Z. Wang, Y.J. Cui, J.M. Yao, Influence of operating conditions on flux in stirring dead-end microfiltration. Technol. Water Treat. **33**, 71–74 (2007)
5. X. Ye, J.M. Yao, Z. Wang, X.C. Liu, D.Z. Liu, H. Zhang, Z.H. Li, Application of linear multi-regression model for specific resistance study in the dead-end microfiltration. J. Beijing Univ. Technol. **33**, 1193–1198 (2007)
6. Z. Wang, Y. Cui, J. Yao, J. Chu, Y. Liang, The influence of various operating conditions on specific cake resistance in the cross-flow microfiltration of yeast suspensions. Desalin. Water Treat. **1**, 237–247 (2009)
7. Z. Wang, J.S. Chu, X.M. Zhang, Study of a cake model during stirred dead-end microfiltration. Deasalination **217**, 127–138 (2007)
8. Z. Wang, J. Chu, Y. Song, Y. Cui, H. Zhang, X. Zhao, Z. Li, J. Yao, Influence of operating conditions on the efficiency of domestic wastewater treatment in membrane bioreactors. Desalination **245**, 73–81 (2009)
9. M. Liu, Z. Wang, L. Shi, Y. Song, W. Dong, Y. Zhou, J. Zhang, L. Yang, Q. Zhang, The influence of operating conditions on the filtration behavior of actual extracellular polymeric substances (EPS) using dead-end membrane filtration cell. Desalin. Water Treat. **44**, 52–59 (2012)
10. L. Yang, Z. Wang, Y. Sun, Z. Hu, S. Zhao, X. Wang, W. Li, X. Xi, J. Zhang, Z. Pei, Influence of various operating conditions on cleaning efficiency in sequencing batch reactor (SBR) activated sludge process. Part II: backwash and water rinsing introduced membrane filtration process. Desalination **272**, 76–84 (2011)
11. Z. Wang, S. Zhao, F. Liu, L. Yang, Y. Song, X. Wang, X. Xi, Influence of operating conditions on cleaning efficiency in sequencing batch reactor (SBR) activated sludge process—Water rinsing introduced membrane filtration process. Desalination **259**, 235–242 (2010)
12. W.J. Wu, Z. Wang, J.M. Yao, S.L. Ji, D.Z. Liu, Membrane fouling and resistance analysis of hollow fiber microfiltration process. Tech. Equip. Environ. Pollut. Control **7**, 30–35 (2006)
13. Z. Wang, X.M. Zhang, W.J. Wu, Influence of operative conditions on permeation flux during dead-end microfiltration. Membr. Sci. Technol. **26**, 26–30 (2006)
14. Y. Zhang, Z. Wang, S.L. Ji, S.Z. Yao, Z.P. Qin, Research on the connection between prepare conditions of the polysulphone base membrane and capability of the PDMS-polysulphone composite membrane. Chem. Res. Appl. **18**, 653–658 (2006)
15. Y.D. Kong, Z. Wang, Y. Ma, L. Hou, W. Yao, Theory investigation on the variation of fouling resistance during water rinsing process of the membrane fouled with sodium alginate. J. Taiwan Inst. Chem. Eng. **66**, 230–238 (2016)
16. C. Zhou, Z. Wang, J.M. Yao, Y. Liu, Optimization of preparing PVDF/CA blend microfiltration membrane by linear multi-regression. J. Chem. Ind. Eng. **58**, 1840–1846 (2007)
17. L. Wang, X. Yuan, Z. Xie, P. Wu, Y. Li, Increasing flash droughts over China during the recent global warming hiatus. *Scientific Reports, 6* (2016)
18. J.M. Yao, Z. Wang, G.M. Sun, J.S. Chu, D.M. Chen, H. Zhang, Z.H. Liu, Prediction of specific resistance for yeast suspensions in dead-end microfiltration. J. Chem. Ind. Eng. **59**, 1430–1435 (2008)
19. X. Xi, Y. Cui, Z. Wang, J. Qian, J. Wang, L. Yang, S. Zhao, Study of dead-end microfiltration features in sequencing batch reactor (SBR) by optimized neural networks. Desalination **272**, 27–35 (2011)

20. Z. Jia, J. Ma, F. Wang, W. Liu, Hybrid of simulated annealing and SVM for hydraulic valve characteristics prediction. Expert Syst. Appl. **38**, 8030–8036 (2011)
21. X. Xi, Z. Wang, J. Zhang, Y. Zhou, N. Chen, L. Shi, W. Dong, L. Cheng, W. Yang, Prediction of impacts of fabrication conditions on the filtration performance of homemade VC-co-VAc-OH microfiltration membrane by support vector machine (SVM). Desalin. Water Treat. **51**, 3970–3978 (2013)
22. K. Gao, X. Xi, Z. Wang, Y. Ma, S. Chen, X. Ye, Y. Li, Use of support vector machine model to predict membrane permeate flux. Desalin. Water Treat. **57**, 16810–16821 (2016)
23. A.S. Jonsson, B. Jonsson, H. Byhlin, A concentration polarization model for the ultrafiltration of nonionic surfactants. J. Colloid Interface Sci. **304**, 191–199 (2006)
24. A.L. Zydney, C.K. Colton, A concentration polarization mode1 for the filtrate flux in cross-flow microfiltration of particulate suspensions. Chem. Eng. Commun. **47**, 1–21 (1986)
25. G. Belfort, R.H. Davis, A.L. Zydney, J. Mem. Sci. **96**, 1–58 (1994)
26. R.H. Davis, J.D. Sherwood, A similarity solution for steady-state crossflow microfiltration. Chem. Eng. Sci. **45**, 3204–3209 (1990)
27. J. Hermia, constant pressure blocking filtration laws application to power law non-Newtonian fluids. Trans. Inst. Chem. Eng. **60**, 183–187 (1982)
28. B.F. Ruth, Studies in Filtration III derivation of general filtration equations. Ind. Eng. Chem. **27**, 708–723 (1935)
29. Ван Чжань, Марцулевич Н.А.Флисюк О. М. Об оценке относительной осадка при разделении суспензии на мембранах. ЖПХ (Россия) **65**(9), 2155–2158 (1992) (Z. Wang, N.A. Martsulevich, O.M. Flisyuk, Evaluation of the relative permeability of precipitate in the separation of suspension on the membrane. J. Appl. Chem. (Russia) **65**(9), 2155–2158 (1992)
30. Z. Wang, S.M. Wang, X.M. Zhang, M. Ma, F.Y. Jie, S.F. Liu, Y. Qian, Experimental study of permeation flux perdiction of hollow fiber microfiltration membrane device under static state operation. *Membrane Science and Technology, 22*, 29–31 (2002)
31. Z. Wang, D. Liu, W. Wu, M. Liu, Study of dead-end microfiltration flux variety law. Desalianation **201**, 175–184 (2006)
32. Z. Wang, L.Y. Yang, Y.W. Lyu, M.N. Aleksandrov, Q. Zhang, D.Z. Liu, X.N. Zhang, Study of dead-end microfiltration flux variety law, Part II: choice of favorable model parameter. Sep. Sci. Technol. **49**, 2657–2667 (2014)
33. C.M. Silva, D.W. Reeve, H. Husain, H.R. Rabie, K.A. Woodhouse, Model for flux prediction in high-shear microfiltration systems. J. Membr. Sci. **173**, 87–98 (2000)
34. Z. Wang, Z.F. Gao, A new flux prediction model for laminar and turbulent flow regimes in constant pressure cross-flow microfiltration. J. Environ. Chem. Eng. **9**, 14 (2021)
35. Z. Wang, Z. Wang, L. Hou, Y. Ma, X. Wang, W. Yao, A corrected cake resistance expression by considering the deposition property of different feed suspensions/solutions. Separat. Sci. Technol. 1–14 (2019)
36. C. Tien, Introduction, in *Principles of Filtration* (2012), pp. 1–6
37. F.M. Tiller, The role of porosity in filtration numerical methods for constant rate and constant pressure filtration based on Kozeny's Law. Chem. Eng. Process **49**, 467–479 (1953)
38. F.M. Tiller, The role of porosity in filtration. II: analytical equations for constant rate filtration. Chem. Eng. Process **51**, 282–290 (1955)
39. F.M. Tiller, The role of porosity in filtration: III Variable pressure-variable rate filtration. AICHE J **4**, 170–174 (1958)
40. F.M. Tiller, H.R. Cooper, The role of porosity in filtration: IV constant pressure filtration. AICHE J. **6**, 595–601 (1960)
41. F.M. Tiller, H.R. Cooper, The role of porosity in filtration: Part V Porosity variation in filter cakes. AICHE J. **8**, 445–449 (1962)
42. F.M. Tiller, M. Shirato, The role of porosity in filtration: VI new definition of filtration resistance. AICHE J. **10**, 61–67 (1964)
43. F.M. Tiller, S.J. Haynes, W.M. Lu, The role of porosity in filtration: VII effect of side-wall friction in compression-permeability cells. AICHE J. **18**, 13–20 (1972)

44. F.M. Tiller, W.M. Lu, The role of porosity in filtration VIII: cake nonuniformity in compression-permeability cells. AICHE J. **18**, 569–572 (1972)
45. F.M. Tiller, T.C. Green, The role of porosity in filtration: IX skin effect with highly compressible materials. AICHE J. **19**, 1266–1269 (1973)
46. F.M. Tiller, C.S. Yeh, The role of porosity in filtration: X deposition of compressible cakes on external radial surfaces. AICHE J. **31**, 1241–1248 (1985)
47. F.M. Tiller, C.S. Yeh, The role of porosity in filtration Part XI: filtration followed by expression. AICHE J. **33**, 1241–1256 (1987)
48. F.M. Tiller, N.B. Hsyung, D.Z. Cong, Role of Porosity in Filtration: XII filtration with sedimentation. AICHE J. **41**, 1153–1164 (1995)
49. F.M. Tiller, J.H. Kwon, Role of porosity in filtration: XIII behavior of highly compactible cakes. AICHE J. **44**, 2159–2167 (1998)
50. C. Tien, S.K. Teoh, R.B.H. Tan, Cake ltration analysis—The effect of the relationship between the poreliquid pressure and the cake compressive stress. Chem. Eng. Sci. **56**, 5361–5369 (2001)
51. Z. Zhu, Z. Wang, H. Wang, Y. Kong, K. Gao, Y. Li, Cake properties as a function of time and location in microfiltration of activated sludge suspension from membrane bioreactors (MBRs). Chem. Eng. J. **302**, 97–110 (2016)
52. S.T. Kelly, A.L. Zydney, Mechanisms for BSA fouling during microfiltration. J. Membr. Sci. **107**, 115–127 (1995)
53. Y.S. Polyakoy, A.L. Zydney, Ultrafiltration membrane performance: effects of pore blockage/constriction. J. Membr. Sci. **434**, 106–120 (2013)
54. Z. Wang, J.S. Chu, W.J. Wu, J.M. Yao, Study of unsteady-state flux prediction in cross-flow microfiltration. Desalination **238**, 290–301 (2009)
55. Z. Wang, Y.J. Cui, W.J. Wu, S.L. Ji, J.M. Yao, H. Zhang, X.Q. Zhao, The convective model of flux prediction in a hollow-fiber module for a steady-state cross-flow microfiltration system. Desalination **238**, 192–209 (2009)
56. Z. Wang, S.L. Ji, X.M. Lu, Y.Y. Gao, Calculation of plated UF device on the stable condition. Technol. Water Treat. **22**, 328–332 (1996)
57. Z. Wang, L.Y. Tan, Z. Zhang, Q.Y. Xi, Calculation of plated UF device under instable condition. J. Beijing Univ. Technol. **24**, 123–128 (1998)
58. D.-J. Chang, F.-C. Hsu, S.-J. Hwang, Steady-state permeate flux of cross-flow microfiltration. J. Membr. Sci. **98**, 97–106 (1995)
59. S.-J. Hwang, D.-J. Chang, C.-H. Chen, Steady state permeate flux for particle cross-flow filtration. Chem. Eng. J. **61**, 171–178 (1996)
60. H. Carrere, F. Blaszkowa, H.R. de Balmam, Modelling the microfiltration of lactic acid fermentation broths and comparison of operating modes. Desalination **145**, 201–206 (2002)
61. A. Bujacz, Structures of bovine, equine and leporine serum albumin. Acta Crystallogr. Sect. D-Biol. Crystallogr. **68**, 1278–1289 (2012)
62. K.A. Majorek, P.J. Porebski, A. Dayal, M.D. Zimmerman, K. Jablonska, A.J. Stewart, M. Chruszcz, W. Minor, Structural and immunologic characterization of bovine, horse, and rabbit serum albumins. Mol. Immunol. **52**, 174–182 (2012)
63. C.C. Ho, A.L. Zydney, A combined pore blockage and cake filtration model for protein fouling during microfiltration. J. Colloid Interface Sci. **232**, 389–399 (2000)
64. G. Bolton, D. Lacasse, R. Kuriyel, Combined models of membrane fouling: development and application to microfiltration and ultrafiltration of biological fluids. J. Membr. Sci. **277**, 75–84 (2006)
65. L. Hou, Z. Wang, P. Song, A precise combined complete blocking and cake filtration model for describing the flux variation in membrane filtration process with BSA solution. J. Membr. Sci. **542**, 186–194 (2017)
66. W. Yao, Z. Wang, P. Song, The cake layer formation in the early stage of filtration in MBR: mechanism and model. J. Membr. Sci. **559**, 75–86 (2018)
67. Q. Li, Z. Wang, C. Liu, X. Wang, A new combined cake-complete model with the cake resistance corrected by cake filtration equilibrium coefficient in cross-flow microfiltration. J. Environ. Chem. Eng. **10**, 11 (2022)

Constant Flowrate Filtration

Nomenclature

A_0	The initial membrane open area (m^2)
$A(t)$	The open membrane area (m^2)
A_a	The apparent membrane area (m^2)
$A_b(t)$	The membrane blocked area (m^2)
$A_s(t)$	The membrane area opened by media scouring (m^2).
B	The particle resuspension rate (s^{-1}); the back-transport coefficient of particles from the membrane surface by shear stress (s^{-1})
C	The volume of particles deposited per unit volume of filtrate
C_{in}	The mass concentration of particles at the inlet of the membrane pores
D	The diffusion coefficient
d	The average pore size of the membrane (m)
J	The permeate flux ($m^3.m^{-2}\ s^{-1}$)
J_0	The initial filtration rate (m/s)
J'	The membrane flux relative to the remaining available membrane area
K	The steady available membrane area ratio
k'	Constant ($kg^{1-n'}m^{n'-2}s^{2n'-2}$)
K_b	Constant (s^{-1})
K_i	The intermediate pore-blocking constant for cross-flow filtration mode
K_m	Coefficient (m^{-3}) ($8\pi nl/A_0{}^2$)
K_s	The standard blocking constant (m^{-1})
$k_2{}'$	Constant ($kg^{1-n_2'}m^{2n_2'-3}s^{n_2'-2}$)
$k_1 = \frac{(1-n_1)\mu\alpha_1\rho sJ}{1-ms}$	Constant ($kg^{1-n}{}_1m^n{}_1{}^{-2}s^{2n}{}_1{}^{-2}$)
K_b	Blocking constant for complete blocking mechanism(s^{-1}); the complete pore-blocking constant (s^{-1})

Z. Wang (ed.), *Modelling of the Microfiltration Process*,
https://doi.org/10.1007/978-981-99-0471-6_3

K_c	The cake filtration constant for cross-flow filtration (m^{-1})
K_{br}	Blocking constant for complete blocking mechanism in constant flowrate filtration (kg^{-1} s^2)
k_c	The proportional constant
L	The pore length (i.e., membrane thickness) (m)
L_m	The length of the membrane pore
l	The equivalent length of the membrane pore (m)
m	Ratio of mass of wet to mass of dry cake
$m_c(t)$	Cake mass (kg)
$m_{att}(t)$	Attached cake mass (kg)
$m_{det}(t)$	Detached cake mass (kg)
N	The number of membrane pores
$N(t)$	The number of the unblocked membrane pores at filtration time t
N_m	The pore density in the membrane
n	The blocking index which can be used to define the filtration mechanism for the complete pore blocking
n_1	The compressibility coefficient of the cake
N_0'	The number of open pores per unit effective membrane area at filtration time t
N_0	The number of open pores per unit effective membrane area for a virgin membrane
$P(t)$	The total trans-membrane pressure (Pa)
$P_c(t)$	The pressure drops across cake (Pa) (originating from the resistance of the deposited cake layer)
P_0	The initial pressure drops across the filter (Pa)
$P_m(t)$	The pressure drops across the membrane (originating from the pore blocking/shrinking resistance of the membrane and the inherent membrane resistance) (Pa)
P_L	The trans-membrane pressure (Pa)
$P_{p,b}$	The pressure drop caused by pore blocking (kPa)
$P_{p,in}$	The pressure drop caused by pore shrinking (kPa)
P^*	The pressure drops across the unfouled region of the filter (Pa)
p	The trans-membrane pressure drops (Pa)
ΔP_c	The pressure across the cake (Pa)
ΔP_m	The pressure across the membrane (Pa)
$Q(t)$	The volumetric flowrate at filtration time t ($m^3 \cdot s^{-1}$)
r	The overall filter mass transfer resistance (m^{-1})
R_0	The virgin membrane resistance (m^{-1})
$R_c(t)$	The cake resistance (m^{-1})
$R_s(t)$	The standard pore-blocking resistance (m^{-1})
$R_t(t)$	The total resistance (m^{-1})
$R_p(t)$	The pore resistance (m^{-1})
R_m	The membrane resistance (m^{-1})
R	The pore radius (m)
$r(t)$	The pore radius at filtration time t(m)

S	The rate of erosion of cake per unit area (kg/(m^{2}s), which is assumed to be invariant with time; soluble microbial products concentration (kg m^{-3})
s	Mass fraction; the mass fraction of solids in the fouling solution
t	The filtration time(s)
u	Flow velocity in membrane pore (m/s)
V	The filtrate volume per unit effective membrane area (m^3/m^2)
V'	The permeate volume relative to the remaining available membrane area
v	Permeate volume per unit effective membrane area (m)
v_m	Fictitious permeate volume per unit membrane area required to obtain cake with flow resistance equivalent to that of membrane (m)
W	The cake mass (kg)
X	Suspended solids concentration in bulk (kg m^{-3})
x'	The number of pores blocked per unit filtrate volume
ω	Parameter of pore-blocking mitigation which is associated with scouring efficiency on pore-blocking mitigation

Greek Symbols

α	The specific cake resistance (m/kg)
α_0	The average specific cake resistance at null stress (m/kg)
α_1	Constant (kg$^{1-n}{}_1$m$^{1+n}{}_1$s$^{2n}{}_1$)
α_{av}	Average specific cake resistance (m/kg)
β	A constant that describes the rate of pore blockage; the scouring parameter (m^2 g^{-1})
β_s	The collision probability between a particle and the membrane surface
β_p	The collision probability between a particle and a membrane pore
β_{pp}	The collision probability between a particle and another particle
δ	Pore-blocking parameter (m^2/g)
γ	The filtrate density (kg/m^3)
θ	The cone angle
μ	Viscosity of permeate (Pa s)
v	Volumetric throughput (volume filtered per unit membrane area)
ρ	Density of filtrate (kg/m^3)
σ	The blocked membrane surface area per unit permeates volume (m^{-1}); soluble microbial product (SMP) fraction rejected by the membrane
τ	The average consolidation time constant
Ω	The rotational Couette speed

1 Introduction

The world is faced with the problem of water shortage. The fundamental measures to solve this problem are to recycle urban sewage and reuse industrial wastewater. In recent years, with the continuous improvement of membrane production technology and the gradual reduction of production cost, the membrane bioreactor (MBR) technology in membrane separation technology has been applied increasingly in municipal wastewater treatment and industrial wastewater reuse. It is because compared with many traditional biological water treatment processes, MBR has high-quality and stable effluent quality, less excess sludge output, a small floor area, and is not limited by the setting occasions. In addition, it can remove ammonia nitrogen and refractory organics and has convenient operation and management advantages, easy-to-realize automatic control, and easy to reform from traditional processes.

Constant flowrate filtration is widely used in MBRs, and the production rate of permeate volume can quickly meet the demand. However, with the filtration process going on, the membrane fouling (Fig. 1), i.e., the accumulation of trapped substances in the membrane pore and membrane surface, and the cake formation on the membrane will increase the trans-membrane pressure (TMP). At this time, monitoring the trans-membrane pressure change required to maintain a stable infiltration rate can better understand the membrane pollution problem in the membrane filtration process. Therefore, establishing the mathematical model of TMP change in the constant flow process helps predict the change of TMP in the filtration process and lays the foundation for the subsequent scientific membrane cleaning.

In the constant flow rate process, the operating mode is also divided into dead-end filtration mode (Fig. 2(a)) and cross-flow filtration mode (Fig. 2(b)). In cross-flow filtration mode, the formation of the fouling cake on the membrane surface was controlled by shear stress generated by fluid flowing on the membrane surface, which is the primary reverse transport mechanism that causes particles to migrate away from the membrane surface (Fig. 2).

(a) (b)

Fig. 1 The membrane bioreactor (MBR) before fouling (**a**) and after fouling (**b**)

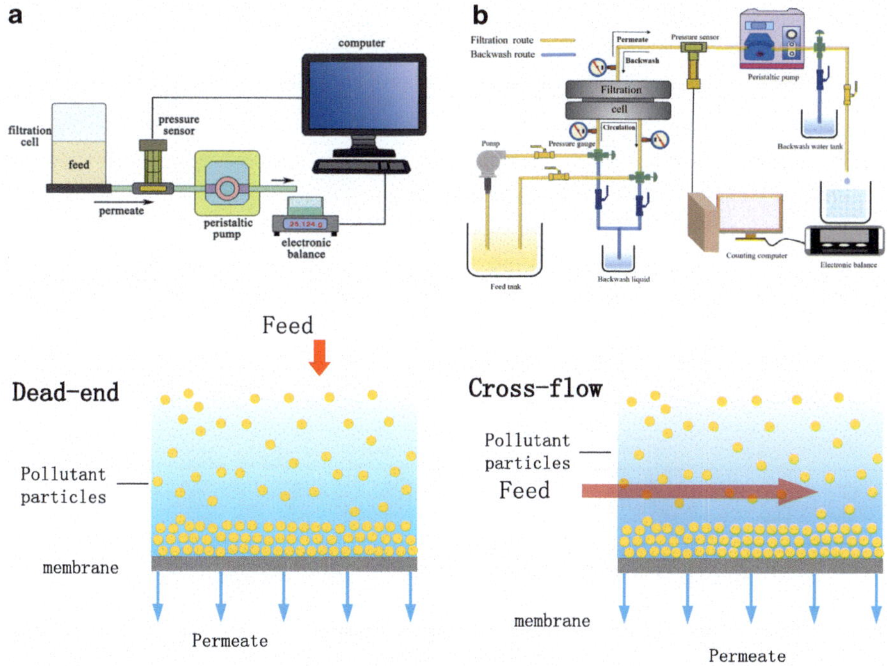

Fig. 2 Schematic diagram of the experimental device. **a** Dead-end device and **b** cross-flow device. Large (black) spheres represent rejected particles, small (gray) spheres represent non-rejected particles, and the arrows indicate the flow direction

2 Fouling Mechanism

2.1 Single Mechanism

The fouling mechanism can be divided into four categories [1]:

1. Complete pore blocking, where foulant particles deposit onto open surface areas (completely blocking the pores in the covered area), but do not deposit on top of each other (Fig. 3(a)).
2. Standard pore blocking, assuming that blocking occurs only inside the pores, where deposited particles reduce the pore diameter (Fig. 3(b)) [2].
3. Intermediate pore blocking is similar to complete pore blocking, but particles are permitted to deposit on top of each other (Fig. 3(c)). Each foulant has a probability to either deposit on an open area of the membrane or deposit onto a previously deposited foulant particle.
4. Cake filtration, foulants completely cover a membrane surface in several layers (Fig. 3(d)). In this case, the cake covers the entire membrane surface area, and the overall filter mass transfer resistance increases proportionally to the cake layer thickness.

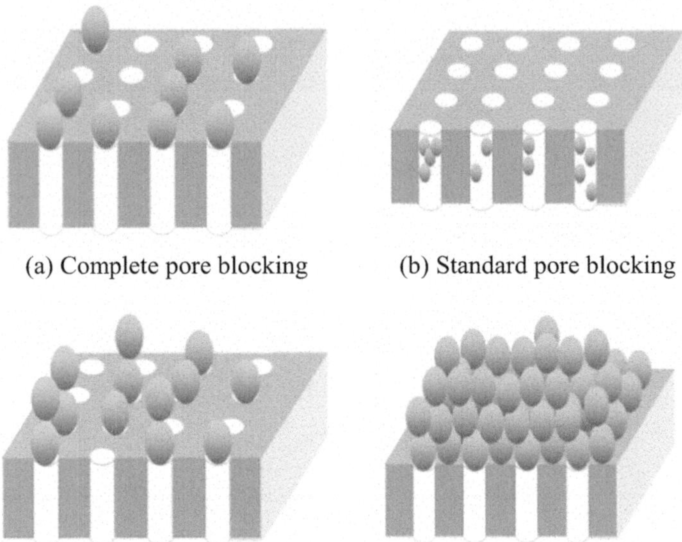

(a) Complete pore blocking (b) Standard pore blocking

(c) Intermediate pore blocking (d) Cake filtration

Fig. 3 Schematic representation of Hermia's fouling mechanism

2.2 Single Mechanism Model in Dead-End Mode for Newtonian Fluid

In the fouling mechanism of the complete blocking, the membrane filtration flux (J) is proportional to the number of open pores of the membrane, and it can be written as

$$J = \frac{dV}{dt} = k_c p N_0' = k_c p\left(N_0 - x'V\right) \tag{1}$$

where V is the filtrate volume per unit effective membrane area (m^3/m^2); t is the filtration time (s); k_c is the proportional constant; p is the trans-membrane pressure drop (Pa); N_0' is the number of open pores per unit effective membrane area at filtration time t; N_0 is the number of open pores per unit effective membrane area for a virgin membrane; and x' is the number of pores blocked per unit filtrate volume.

Therefore, the pressure can be written as

$$p = \frac{J}{k_c(N_0' - xv)} \tag{2}$$

After differentiating Eq. (2) under constant flowrate ($J = J_O =$ const), the following equation can be obtained:

$$\frac{dp}{dv} = \frac{k_c x}{J_0} p^2 = \frac{K_b}{J_0 p_0} p^2 = K_{br} p^2 \tag{3}$$

where J_0 is the initial filtration rate (m/s), K_b (s^{-1}) is blocking constant for complete blocking mechanism, K_{br} is blocking constant for complete blocking mechanism in constant flowrate filtration ($kg^{-1}\ s^2$), and $K_{br} (= k_c x / J_0)$ is the blocking constant for complete blocking mechanism in constant rate filtration.

Therefore, four blocking filtration laws in the case of constant flowrate filtration can be represented by a general differential equation, i.e., the constant flowrate blocking laws [2–4]:

$$\frac{dp}{dv} = k' p^{n'} \tag{4}$$

where k' is constant ($kg^{1-n'} m^{n'-2} s^{2n'-2}$) and n' is constant.

A power index n' is the blocking index which can be used to define the filtration mechanism for the complete pore blocking; the standard pore blocking; the intermediate pore blocking; and the cake filtration at n' of 2, 1.5, 1, and 0, respectively.

The equations derived for constant flowrate filtration are presented in Table 1. The corresponding graphical identification of the constant flowrate blocking laws is illustrated in Fig. 3, which can be served as the judgment of the blocking filtration law. The filtrate volume v per unit membrane area is directly proportional to the filtration time t in constant flowrate filtration process, i.e., $v = J_0 t$. Consequently, it is also possible to identify the dominant blocking filtration law by plotting the data against t in place of v as shown in Fig. 4.

(a) Cake filtration ($n' = 0$)

The value of n' is 0 for the cake filtration mechanism in which an incompressible filter cake forms during filtration. However, using Eq. (5) requires considerable attention when the filter cake exhibits a compressible behavior.

$$p = \frac{\mu \alpha_{av} \rho s J (v + v_m)}{(1 - ms)} \tag{5}$$

where α_{av} is average specific cake resistance (m/kg), μ is viscosity of permeate (Pa s), ρ is density of filtrate (kg/m^3), s is mass fraction, v_m is fictitious permeate volume

Table 1 Blocking filtration equations for constant flowrate filtration

Function	Complete blocking	Standard blocking	Intermediate blocking	Cake filtration
$\frac{dp}{dv} = k' p^{n'}$	$n' = 2.0$	$n' = 1.5$	$n' = 1.0$	$n' = 0$
$p = f(v)$	$\frac{p_0}{p} = 1 - \frac{K_b}{J_0} v$	$\left(\frac{p_0}{p}\right)^{1/2} = 1 - \frac{K_s}{2} v$	$ln\left(\frac{p_0}{p}\right) = K_i v$	$\frac{p_0}{p} = 1 + K_c J_0 v$

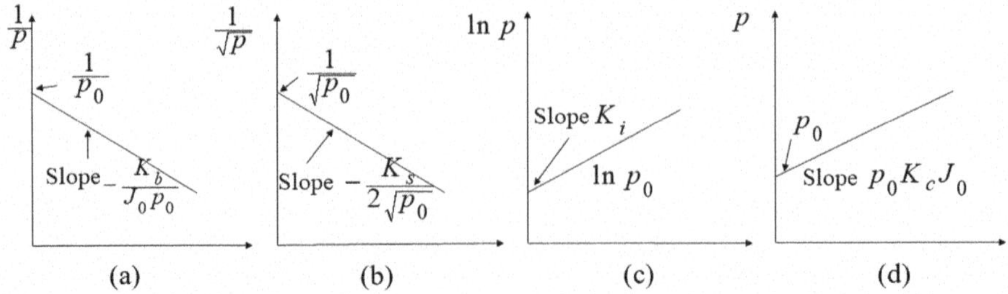

Fig. 4 Graphical identification of blocking filtration laws for constant rate filtration: **a** complete blocking law, **b** standard blocking law, **c** intermediate blocking law, and **d** cake filtration law

per unit membrane area required to obtain cake with flow resistance equivalent to that of membrane (m), m is ratio of mass of wet to mass of dry cake, and v is permeate volume per unit effective membrane area (m).

The average specific cake resistance α_{av} in Eq. (5) is related to the applied filtration pressure p [5, 6].

$$\alpha_{av} = \alpha_1 p^{n'} \tag{6}$$

where α_1 is constant ($kg^{1-n}{}_1 m^{1+n}{}_1 s^{2n}{}_1$) and n_1 is constant, and n_1 is especially termed the compressibility coefficient of the cake. The higher the n_1 value indicates higher the compressibility of the cake.

After substituting Eq. (6) into Eq. (5):

$$p^{1-n_1} = \frac{\mu\alpha_1\rho s J}{(1-ms)}(v+v_m) \tag{7}$$

After differentiation of Eq. (7) for to v under the constant flowrate:

$$\frac{dp}{dv} = \frac{(1-n_1)\mu\alpha_1\rho s J}{1-ms}p^{n_1} = k_1 p^{n_1} \tag{8}$$

where $k_1 = \frac{(1-n_1)\mu\alpha_1\rho s J}{1-ms}$ ($kg^{1-n}{}_1 m^{n}{}_1{}^{-2} s^{2n}{}_1{}^{-2}$) is a constant.

Therefore, the value of n' in Eq. (4) is not 0 for constant flowrate cake filtration in which a compressible cake forms in the filtration process and it increases as the compressible behaviors become more pronounced.

(b) Intermediate pore blocking ($n' = 1.0$)

Suppose the compressibility coefficient n' is 1.0, as seen in a highly compressible cake. In that case, the power index n' in the differential equation for the cake filtration shows the same value as that derived for the intermediate blocking law, thereby leading to serious confusion. Since v is directly proportional to the time t in constant flowrate filtration process, Eq. (4) is rewritten as

$$\frac{dp}{dv} = k_2 p^{n'} \tag{9}$$

where k_2 is a constant, it should be noted that the characteristic differential equation can also be represented in the following form:

$$\frac{dt}{dp} = \frac{1}{k_1 p^{n'}} \tag{10}$$

$$\frac{d(dt/dp)}{dp} = \frac{1}{k_1}\frac{d(p^{n'})}{dp}, \frac{d^2 t}{dp^2} = k_1'\left(\frac{dt}{dp}\right)^{n_1'} \tag{11}$$

where k_1' and n_1' are constants. Equation (10) is similar to equation $\frac{d^2t}{dv^2} = k(\frac{dt}{dv})^n$ in the form, by considering that p in Eq. (10) corresponds to v in Eq. $\frac{d^2t}{dv^2} = k(\frac{dt}{dv})^n$.

Alternatively, the characteristic form can be expressed as [3, 4]

$$\frac{d^2 p}{dv^2} = k_2'\left(\frac{dp}{dv}\right)^{n_2'} \tag{12}$$

where $k_2'\left(kg^{1-n_2'} m_2^{2n_2'-3} s_2^{n_2'-2}\right)$ and n_2' are constants. Also, in this case, using Eq. (12) requires a considerable attention in applying it to filtration, forming a compressible cake.

(c) Standard pore blocking (n = 1.5)

The expression of the average specific cake resistance (α_{av}) is increasing with pressure (p) in the analysis of constant flowrate filtration behaviors, as described by [3, 7]

$$\alpha_{av} = \alpha_0 + \alpha_2 p \tag{13}$$

where α_0 (m/kg) and α_2 (kg^{-2} m^2 s^2) are constants, and α_0 is the average specific cake resistance at null stress. Equation (13) has been applied to microbial suspensions, which exhibit highly compressible behaviors under relatively high-pressure conditions. In this case, the value of exponent n_2' in Eq. (12) becomes 1.5 and thus is the same as the complete blocking law:

$$\left(\frac{p_0}{p}\right)^{1/2} = 1 - \frac{K_s}{2} v \tag{14}$$

(d) Complete pore blocking (n = 2.0)

The relation among the filtration rate J, applied filtration pressure p, and filtrate volume v per unit membrane area can be obtained based on Eq. (1) and thus it is represented by [4, 8]

$$\frac{J}{p} = \frac{J_0}{p_0} - \frac{K_b}{p_0} v \tag{15}$$

$$\frac{p_0}{p} = 1 - \frac{K_b}{J_0} v \tag{16}$$

2.3 Single Mechanism Model in Dead-End Mode (Non-Newtonian Fluids)

The rheological equation representing the relation between the shear stress τ and shear rate $\dot{\gamma}$ for power-law non-Newtonian fluids can be written as follows:

$$\tau = k\gamma^n \tag{17}$$

where k is the fluid consistency index and n is the fluid behavior index representing the intensity of non-Newtonian behaviors. n is 1 for Newtonian fluid, while n departs from 1.0 for non-Newtonian fluid.

The relationship between the average flow rate u and the tube radius r for power-law non-Newtonian fluids can be expressed as [9]

$$u = \frac{N}{3N+1}\left(\frac{p}{2KL}\right)^{1/N} r^{(N+1)/N} \tag{18}$$

The characteristic form of constant flowrate blocking laws for non-Newtonian flow can be expressed as [1]

$$\frac{dp}{dv} = k'_N p^{n'_N} \tag{19}$$

Therefore, the fouling mechanism depended on the value of the power index (Table 2):$n'_N = \frac{N+1}{N}$ is complete pore blocking, when the $n'_N = \frac{3N+3}{3N+1}$ is standard pore blocking, $n'_N = 1$ is intermediate pore blocking, and $n_N{}' = 0$ is the cake filtration..

2.4 Single Mechanism Model in Cross-Flow Mode

(a) Complete pore blocking

According to Darcy's law [10]:

Table 2 Blocking filtration equations for constant rate filtration for power law (non-Newtonian fluids)

Function	Complete blocking	Standard blocking	Intermediate blocking	Cake filtration
$\frac{dp}{dv} = k'_N p^{n'_N}$	$n'_N = \frac{N+1}{N}$	$n'_N = \frac{3N+3}{3N+1}$	$n'_N = 1$	$n'_N = 0$
$p = f(v)$	$\left(\frac{p_0}{p}\right)^{1/N} = 1 - \frac{K_b}{J_b} v$	$\left(\frac{p_0}{p}\right)^{2/(3N+1)} = 1 - \frac{K_s}{2} v$	$\ln\left(\frac{p}{p_0}\right) = N K_i v$	$\frac{p}{p_0} = K_c J_0^N v + 1$

Note n′ is the power index for power-law non-Newtonian fluids-solids mixtures and k′ N is combination constant ($kg^{1-n' N} \cdot m\, n'^{N-2} \cdot s^2\, n'^{N-2}$)

$$Q = \frac{A \Delta P}{\mu R} \tag{20}$$

where R is the overall filter mass transfer resistance (m^{-1}).

For constant flowrate mode, Q_0 ($t = 0$) should equal Q_t ($t = t$). Following Hermia's approach, the overall filter resistance is constant, attributing the increase in ΔP in the filtration process to a reduction in the available membrane surface area. Therefore, comparing the expressions for Q_0 and Q_t from Eq. (20):

$$\Delta P_t = \frac{\Delta P_0 A_0}{A_t} \tag{21}$$

where the subscripts *0* and *t* reflect initial condition and condition at time *t*, respectively.

The available membrane surface area in complete pore blocking is a function of the filtered volume of solution [1, 4, 11]:

$$A_t = A_0 - \sigma V \tag{22}$$

where V is the permeate volume that has passed through the membrane (m^3) and σ is the blocked membrane surface area per unit permeate volume (m^{-1}).

In constant flowrate mode, the permeate flowrate is constant and V can be expressed as the product of the initial flowrate and the filtration time:

$$V = Q_0 t = A_0 J t \tag{23}$$

Substituting Eq. (23) into Eq. (22):

$$A_t = A_0 - \sigma A_0 J t \text{ and } -\frac{dA}{dt} = \sigma A_0 J \tag{24}$$

Based on Field et al. [11], when accounting for cross flow, the rate of change in the available membrane surface area is a function of two counteracting processes:

foulant deposition, and foulant removal [11, 12]. Foulant removal is taken to be proportional to the blocked filtration area. If particles are removed at a constant rate from blocked pores, the decrease in the available membrane surface area is given by [13]:

$$-\frac{dA}{dt} = \sigma A_0 J - B(A_0 - A) \tag{25}$$

where B is the particle resuspension rate (s^{-1}), a constant reflecting the frequency at which foulants are removed from the membrane surface by the cross-flow shear force. The first term on the right-hand side of Eq. (25) is the foulant deposition term, and the second is the foulant removal term.

Integration of Eq. (25) yields an expression for the available membrane surface area as a function of time:

$$A_t = A_0(1 - \frac{\sigma J}{B}(1 - \exp(-Bt))) \tag{26}$$

After substituting Eq. (26) into Eq. (21), the time dependence of ΔP for complete pore blocking can be obtained as follows:

$$\Delta P_t = \frac{\Delta P_0}{(1 - \frac{\sigma J}{B}(1 - \exp(-Bt)))} \tag{27}$$

(b) Intermediate pore blocking

Following Hermia's development, the probability of a particle depositing onto an open pore rather than on a previously deposited particle is proportional to the instantaneous unobstructed membrane surface area [11, 14]. The rate of change of available membrane surface area, based on Eq. (24), is expressed by

$$-\frac{dA}{dt} = \sigma AJ \tag{28}$$

The only difference between Eqs. (28) and (24) is the area accounted for on the right-hand side of the equation. In intermediate pore blocking, each foulant particle has a probability to either block an open pore or deposit on a previously deposited foulant particle [14]. The probability of blocking open pores decreases as additional pores are blocked. Therefore, to account for the decreasing blocking rate during fouling, $\sigma A_0 J$ from Eq. (24) is changed to σAJ.

In cross-flow filtration process, the foulant removal term is identical to that introduced for complete pore blocking (Eq. 25):

$$-\frac{dA}{dt} = \sigma AJ - B(A_0 - A) \tag{29}$$

After integration of Eq. (29)

$$A_t = A_0\left(\frac{1}{K_i} + \left(1 - \frac{1}{K_i}\right)\exp(-K_i Bt)\right) \tag{30}$$

where K_i is the intermediate pore-blocking constant for cross-flow filtration mode and is defined as

$$K_i = \frac{B + \sigma J}{B} \tag{31}$$

After substitution of Eq. (30) into Eq. (21), the time dependence of ΔP for intermediate pore blocking can be obtained as follows:

$$\Delta P_t = \frac{\Delta P_0}{(\frac{1}{K_i} + (1 - \frac{1}{K_i})\exp(-K_i Bt))} \tag{32}$$

(c) Standard pore blocking

In contrast to the previously described fouling mechanisms, standard pore blocking assumes fouling occurs inside the membrane pores rather than on the surface [3].

Laminar flow through a straight cylindrical pore can be described using the Hagen-Poiseuille equation [14]:

$$Q = N\left(\frac{\pi r^4 \Delta P}{8\mu L}\right) \tag{33}$$

where N is the number of membrane pores, r is the pore radius (m), and L is the pore length (i.e., membrane thickness) (m).

If the initial and final flowrates are equal, the relationship of ΔP_t to ΔP_0 is a function of the initial radius, r_0, and the radius at time t, r_t:

$$\Delta P_t = \left(\frac{r_0}{r_t}\right)^4 \Delta P_0 \tag{34}$$

Following Hermia, a solids mass balance yields [1, 3]

$$N\pi L\left(r_0^2 - r_t^2\right) = CV \tag{35}$$

where C is the volume of particles deposited per unit volume of filtrate.

Rearranging Eq. (35) yields the desired ratio:

$$\left(\frac{r_t}{r_0}\right)^4 = \left(1 - \frac{CV}{N\pi L r_0^2}\right)^2 \tag{36}$$

Defining K_s as the constant for standard pore blocking (m^{-3}):

$$K_s = \frac{C}{N\pi L r_0^2} \tag{37}$$

After substitution of Eq. (36) into Eq. (34), the time dependence of ΔP for standard pore blocking can be obtained as

$$\Delta P_t = \frac{\Delta P_0}{(1 - K_s A_0 J t)^2} \tag{38}$$

In addition, a model assumes that fouling occurs only near the entrance of the filter, with the rest of the filter remaining unaffected by the glycoconjugate. The total pressure drop through the filter is thus

$$P = (P_0 - P^*)(\frac{N_0}{N}) + P^* \tag{39}$$

where N is number of membrane pores.

The pressure drop across the fouled region increases with time due to pore blockage, with the rate of change in the number of open pores evaluated as

$$\frac{d\left(\frac{N}{N_0}\right)}{dt} = -\beta J\left(\frac{N}{N_0}\right) \tag{40}$$

Equation (40) is consistent with the form used in the classical intermediate pore blockage model [15]. The trans-membrane pressure can be evaluated as a function of the volumetric throughput by integrating Eq. (40) to give [13]

$$P = (P_0 - P^*)\exp(\beta v) + P^* \tag{41}$$

where P_0 is the initial pressure drop across the filter, P^* is the pressure drop across the unfouled region of the filter (which remains constant during filtration at constant filtrate flux), β is a constant that describes the rate of pore blockage, and v is volumetric throughput (volume filtered per unit membrane area).

Equation (43) reduces to the intermediate pore blockage model when $P^* = 0$, i.e., when the fouling blocks the flow through the entire filter depth.

In order to explore the fouling behavior in more detail, Eq. (40) was used to evaluate the normalized pressure gradient:

$$\frac{1}{P}\left(\frac{dP}{dt}\right) = \beta\left(1 - \frac{P^*}{P}\right) \tag{42}$$

In each case, the derivatives were evaluated numerically directly from the pressure data using appropriate difference representations.

(d) Cake filtration

In cake filtration, the overall filter resistance is expressed as the sum of the clean membrane resistance and the resistance of the deposited foulant layers. In addition, an erosion term is added to account for the removal of foulant layers due to cross flow [11]:

$$R_t = R_0 + \frac{\alpha W}{a_0} - \alpha St \tag{43}$$

where α is the specific cake resistance (m/kg), W is the cake mass (kg), and S is the rate of erosion of cake per unit area (kg/(m^{2}s), which is assumed to be invariant with time.

From a mass balance on the cake [14, 16]:

$$W = \frac{V\gamma s}{(1 - ms)} \tag{44}$$

where γ is the filtrate density (kg/m^3), s is the mass fraction of solids in the fouling solution, and m is the mass ratio of wet-to-dry cake.

After substitution of Eq. (44) into Eq. (43) results in an expression for overall filter resistance as a function of time:

$$R_t = R_0(1 + K_c Jt) \tag{45}$$

where K_c is the cake filtration constant for cross-flow filtration (m^{-1}):

$$K_c = \frac{\alpha\gamma s}{R_0(1 - ms)} - \frac{\alpha S}{J R_0} \tag{46}$$

After substitution of Eq. (45) into Darcy's Law, the time dependence of ΔP for cake filtration can be obtained as

$$\Delta P_t = \Delta P_0(1 + K_c Jt) \tag{47}$$

Equation (47) predicts a linear rise in ΔP with filtration time. The assumption that the cake covers the entire membrane surface area cannot be strictly true at short filtration time before a cake layer has deposited. The time required for deposition of the first layer of foulants depends on the permeate flux. When the membrane surface has been covered, cake filtration is expected to be the dominant fouling mechanism at long filtration times (Table 3).

Table 3 Summary of the relationships between ΔP and filtration time for the four proposed fouling mechanisms in constant permeate flux cross-flow filtration

Complete	Intermediate	Cake	Standard
$\Delta P_t = \frac{\Delta P_0}{(1-\frac{\sigma J}{B}(1-\exp(-Bt))}$	$\Delta P_t = \frac{\Delta P_0}{(\frac{1}{K_i}+(1-\frac{1}{K_i})\exp(-K_i Bt))}$	$\Delta P_t = \Delta P_0(1 + K_c Jt)$	$\Delta P_t = \frac{\Delta P_0}{(1-K_s a_0 Jt)^2}$

2.5 Combined Two-Mechanism Model

(1) Standard blocking model

In the standard model, membranes have straight cylindrical pores that decline in radius as solid matter accumulates on the pore walls [17] and the resistance can be calculated as a function of time:

$$R = R_0\left(1 + \frac{K_s J_0 t}{2}\right)^2 \tag{48}$$

where J_0 is the initial flux (m/s) and K_s is the standard blocking constant (m^{-1}).

(2) Complete and intermediate blocking models

The available membrane area declines with permeate volume for the complete blocking mechanism [17]:

$$\frac{A}{A_0} = 1 - \frac{K_b}{J_0}V \tag{49}$$

The available membrane area declines with permeate volume in the intermediate blocking mechanism [17]:

$$\frac{A}{A_0} = \exp(-K_i V) \tag{50}$$

The constant K_b has units of s^{-1} and K_i has units of m^{-1}. These equations are valid for both constant pressure and constant flow operation mode.

(3) Cake filtration model

In the cake filtration mechanism, the resistance to flow is increased by the presence of a cake layer on the membrane surface. Therefore, the total resistance R will be the sum of the membrane resistance and the cake resistance. R can be calculated as a function of time:

$$\frac{R}{R_0} = (1 + 2K_c J_0^2 t)^{1/2} \tag{51}$$

2.5.1 Combined Two-Mechanism Models

1. Bolton's model [17]

① Complete pore-blocking and cake filtration model

In the constant flowrate filtration process, both the pressure and the resistance increased due to pore blocking and the cake formation on the membrane surface:

$$\frac{P}{P_0} = \frac{RA_0}{R_0 A} \tag{52}$$

From the complete pore-blocking mechanism (Table 4):

$$\frac{A}{A_0} = 1 - K_b t \tag{53}$$

After substitution into Eq. (52):

$$J' = \frac{J_0}{1 - K_b t} \tag{54}$$

where J' is the membrane flux relative to the remaining available membrane area.
After integration of Eq. (54):

$$V' = \frac{-J_0}{K_b} \ln(1 - K_b t) \tag{55}$$

where V' is the permeate volume relative to the remaining available membrane area.
From the cake filtration mechanism (Table 4):

$$\frac{R}{R_0} = \left(1 + K_c J_0 V'\right) \tag{56}$$

After combining Eqs. (55) and (56):

$$\frac{R}{R_0} = \left(1 - \frac{K_c J_0^2}{K_b} \ln(1 - K_b t)\right) \tag{57}$$

Substituting Eqs. (53) and (57) into Eq. (52) gives

$$\frac{P}{P_0} = \frac{R}{R_0} \cdot \frac{A_0}{A} = \frac{1}{(1 - K_b t)} \left(1 - \frac{K_c J_0^2}{K_b} \text{In}(1 - K_b t)\right) \tag{58}$$

Table 4 Four single fouling mechanisms to derive combined models

Fouling mechanism	Pressure variation	–	–	–
Complete blocking	$\frac{p_0}{p} = 1 - \frac{K_b}{J_0} v$	$\frac{R}{R_0} = 1$	$\frac{A}{A_0} = 1 - \frac{K_b}{J_0} V$	$\frac{A}{A_0} = 1 - K_b t$
Intermediate blocking	$In\left(\frac{p}{p_0}\right) = K_i v$	$\frac{p}{p_0} = \frac{RA_0}{R_0 A}$	$\frac{A}{A_0} = e^{-K_i V}$	$\frac{A}{A_0} = e^{-K_i J_0 t}$
Standard blocking	$\left(\frac{p_0}{p}\right)^{1/2} = 1 - \frac{K_s}{2} v$	$\frac{p}{p_0} = \frac{RA_0}{R_0 A}$	$R = R_0\left(1 - \frac{K_s V}{2}\right)^{-2}$	$R = R_0\left(1 - \frac{K_s J_0 t}{2}\right)^{-2}$
Cake filtration	$\frac{p_0}{p} = 1 + K_C J_0 v$	$\frac{A}{A_0} = 1$	$\frac{R}{R_0} = (1 + K_c J_0 V)$	$\frac{R}{R_0} = \left(1 + K_c J_0^2 t\right)$

② Intermediate pore-blocking and cake filtration model

From the intermediate pore-blocking mechanism (Table 4):

$$\frac{A}{A_0} = \exp\ (-K_i J_0 t) \tag{59}$$

After substitution into Eq. (52)

$$J' = \frac{J_0}{\exp\ (-K_i J_0 t)} \tag{60}$$

After integration of Eq. (60):

$$V' = \frac{1}{K_i}(\exp(K_i J_0 t) - 1) \tag{61}$$

From the cake filtration mechanism (Table 4):

$$\frac{R}{R_0} = \left(1 + K_c J_0 V'\right) \tag{62}$$

After combining Eqs. (61) and (62):

$$\frac{R}{R_0} = \left(1 + K_c J_0 V'\right) = \left(1 + K_c J_0 \frac{1}{K_i}(\exp(K_i J_0 t) - 1)\right) \tag{63}$$

Substituting Eq. (63) and Eq. (53) into Eq. (52) gives

$$\frac{P}{P_0} = \frac{R}{R_0} \cdot \frac{A_0}{A} = \exp\ (K_i J_0 t)\left(1 + \frac{K_c J_0^2}{K_i}(\exp(K_i J_0 t) - 1)\right) \tag{64}$$

③ Standard pore blocking and complete pore-blocking model

From the standard filtration mechanism (Table 4):

$$\frac{R}{R_0} = \left(1 - \frac{K_s V}{2}\right)^{-2} \tag{65}$$

As discussed above, Eq. (55) was obtained for the complete pore-blocking mechanism. If Eqs. (55) and (65) are combined, we get

$$\frac{R}{R_0} = \left(1 + \frac{K_s J_0}{2K_b} In(1 - K_b t)\right)^{-2} \tag{66}$$

Substituting Eq. (66) into Eq. (52) gives

$$\frac{P}{P_0} = \frac{R}{R_0} \cdot \frac{A_0}{A} = \left(1 + \frac{K_s J_0}{2K_b} \text{In}(1 - K_b t)\right)^{-2} \cdot \frac{A_0}{A} \tag{67}$$

If considering the available membrane area declines with time for complete blocking mechanism, i.e., Eq. (53), the following equation can be obtained:

$$\frac{P}{P_0} = \frac{1}{(1 - K_b t)\left(1 + \frac{K_s J_0}{2K_b} In(1 - K_b t)2\right)} \tag{68}$$

④ Intermediate pore blocking and standard pore-blocking model

For intermediate pore-blocking mechanism, Eq. (61) can be obtained after substitution of Eq. (59) into Eq. (52).

From the standard blocking mechanism (Table 4):

$$R = R_0\left(1 - \frac{K_s V}{2}\right)^{-2} \tag{69}$$

After substituting Eq. (61) into Eq. (69), we get

$$R = R_0\left(1 - \frac{K_s}{2K_i} \exp(K_i J_0 t) - 1\right)^2 \tag{70}$$

Pressure as a function of time in the combined standard and intermediate blocking model can be obtained after substituting Eqs. (70) and (59) into Eq. (52):

$$\frac{P}{P_0} = \frac{R}{R_0} \cdot \frac{A_0}{A} = \frac{\exp(K_i J_0 t)}{\left(1 - \frac{K_s}{2K_i} \exp(K_i J_0 t) - 1\right)^2} \tag{71}$$

⑤ Cake filtration and standard pore-blocking model

For the combined cake-standard blocking model, the resistance increases over the course of the run and the membrane area remains constant.

From the cake filtration mechanism (Table 4):

$$\frac{Rc}{R_0} = \frac{R - R_0}{R_0} = \left(1 + K_c J_0 V'\right) \tag{72}$$

From the standard filtration mechanism Table 4 and Eq. (65) can be obtained.

If substituting Eqs. (65) and (72) into Eq. (52) considering the membrane area remains constant and $V = J_0 t$:

Table 5 Summary of the five constant flow combined fouling models

Model	Equation	Fitted parameters
Cake-complete	$\frac{P}{P_0} = \frac{1}{(1-K_b t)}\left(1 - \frac{K_c J_0^2}{K_b} \ln(1 - K_b t)\right)$	$K_c(s/m^2)$, $K_b(s^{-1})$
Cake-intermediate	$\frac{P}{P_0} =$ $\exp(K_i J_0 t)\left(1 + \frac{K_c J_0}{K_i} + \exp(K_i J_0 t) - 1\right)$	$K_c(s/m^2)$, $K_i(m^{-1})$
Complete-standard	$\frac{P}{P_0} = \frac{1}{(1-K_b t)\left(1+\frac{K_s J_0}{2K_b} \ln(1-K_b t)\right)^2}$	$K_b(s^{-2})$, $K_s(m^{-1})$
Intermediate-standard	$\frac{P}{P_0} = \frac{\exp(K_i J_0 t)}{\left(1+\frac{K_s}{2K_i} \exp(K_i J_0 t)-1\right)^2}$	K_i (m^{-1}), $K_s(m^{-1})$
Cake-standard	$\frac{P}{P_0} = \left((1 - \frac{K_s J_0 t}{2})^{-2} + K_c J_0^2 t\right)$	$K_c(s/m^2)$, $K_s(m^{-1})$

$$\frac{P}{P_0} = \frac{R}{R_0} \cdot \frac{A_0}{A} = \frac{R}{R_0} = \left(\left(1 - \frac{K_s J_0 t}{2}\right)^{-2} + K_c J_0^2 t\right) \tag{73}$$

A summary of the five new constant flow combined fouling models, including the names, component mechanisms, equations, and fitted parameters, is provided in Table 5.

2. Charfi's model [18]

Assuming the membrane fouling mechanism is pore blocking and cake formation.

$$\frac{dA(t)}{dt} = A_0 - \frac{dA_b(t)}{dt} + \frac{dA_s(t)}{dt} \tag{74}$$

where $A(t)$ is the open membrane area (m^2), $A_b(t)$ is membrane blocked area (m^2), and $A_s(t)$ is the membrane area opened by media scouring (m^2).

The variation of $A_b(t)$ with time can be expressed by a first-order differential equation:

$$\frac{dA_b(t)}{dt} = A(t)\delta J \sigma S \tag{75}$$

where δ is pore-blocking parameter (m^2/g), J is the permeate flux ($m^3.m^{-2}\ s^{-1}$), σ is soluble microbial product (SMP) fraction rejected by the membrane, and S is soluble microbial products concentration (kg m^{-3}).

The transient variation of the membrane area cleaned by continuous scouring effect ($A_s(t)$) was assumed proportional to $A(t)$ and ω

$$\frac{dA_s(t)}{dt} = A(t)\omega \tag{76}$$

where ω is parameter of pore-blocking mitigation which is associated with scouring efficiency on pore-blocking mitigation.

In constant flowrate mode, TMP_p due to pore-blocking mechanism can be quantified as

$$TMP_{pore} = \frac{J A_0 R_0 \mu}{A(t)} \tag{77}$$

where J is the permeate flux ($m^3.m^{-2}\ s^{-1}$), R_0 is the intrinsic membrane resistance (m^{-1}), μ is the permeate viscosity (Pa.s), A_0 is the initial membrane open area (m^2), and A(t) is the membrane open area at a time t.

Pressure variation caused by cake formation (TMP_c) can be expressed as

$$TMP_c(t) = \mu J R_c(t) \tag{78}$$

where J is the permeate flux ($m^3.m^{-2}\ s^{-1}$), μ is the permeate viscosity (Pa.s), and R_c(t) is the cake resistance (m^{-1}).

Mass of the cake on membrane can be expressed as

$$\frac{dm_c(t)}{dt} = \frac{dm_{att(t)}}{dt} - \frac{dm_{det}(t)}{dt} \tag{79}$$

where $m_c(t)$ is cake mass (kg), $m_{att}(t)$ is attached cake mass (kg), and $m_{\ det}\ (t)$ is detached cake mass (kg).

The variation of the foulants' mass driven by convective forces ($m_{att}(t)$) with time can be expressed as

$$\frac{dm_{att}(t)}{dt} = AJ(X + S\sigma) \tag{80}$$

where J is the permeate flux ($m^3.m^{-2}\ s^{-1}$), X is suspended solids concentration in bulk ($kg\ m^{-3}$), S is the SMP concentration in bulk ($kg\ m^{-3}$), and σ is soluble microbial product (SMP) fraction rejected by the membrane.

Assume the mass of contaminant detached from the membrane surface by particle sparging was proportional to (i) the specific flux of contaminants brought to the membrane surface ($\frac{dm_{att}(t)}{dt}$) and to (ii) the specific mass of the contaminants accumulated on the membrane surface to form cake mass ($m_c(t)$). The variation of mass of matters detached from the deposit by particle sparging ($dm_{de\ t}\ (t)$) with time can be expressed as [19, 20]

$$\frac{dm_{det}(t)}{dt} = \beta \frac{dm_{att}(t)}{dt} \frac{m_c(t)}{A(t)} \tag{81}$$

where β is the scouring parameter ($m^2\ g^{-1}$), m_c is the cake specific mass (g), and $A(t)$ is the membrane open area (m^{-2}).

The cake resistance on the membrane surface can be expressed as

$$R_c(t) = \alpha \frac{m_c(t)}{A(t)} \tag{82}$$

where $R_c\ (t)$ is cake resistance (m^{-1}), $m_c\ (t)$ is cake mass (kg), A(t) is the membrane open area, and α is specific cake resistance (m kg^{-1}).

α in Eq. (82) can be expressed as

$$\alpha = \alpha_0 TMP_c^n(t) \tag{83}$$

where TMP_c is trans-membrane pressure due to the cake formation (Pa).

According to Darcy's law, TMP_c in Eq. (82) can be expressed as

$$TMP_c(t) = \mu J R_c(t) \tag{84}$$

where J is the permeate flux ($m^3.m^{-2}\ s^{-1}$), μ is the permeate viscosity (Pa.s), and R_c is the cake resistance (m^{-1}).

By combining Eqs. (80), (83), and (84):

$$TMP_c(t) = \mu J R_c(t) = \mu J \alpha \frac{m_c(t)}{A(t)} = \mu J \frac{m_c(t)}{A(t)} \alpha_0 TMP_c^n(t) \tag{85}$$

$$TMP_c(t) = \left(\mu J \frac{m_c(t)}{A(t)} \alpha_0 \right)^{\frac{1}{1-n}} \tag{86}$$

Therefore, the total TMP variation with time can be expressed as follows:

$$TMP(t) = TMP_c(t) + TMP_{pore}(t) = \left(\mu J \frac{m_c(t)}{A(t)} \alpha_0 \right)^{\frac{1}{1-n}} + \frac{J A_0 R_0 \mu}{A(t)} \tag{87}$$

The proposed model showed good agreement between model productions and experimental data to assess fouling mitigation by the scouring effect of fluidized media in fluidized membrane bioreactor ($R^2 = 0.9923$) (Fig. 5).

(3) Du's model [19]

According to Darcy's law:

$$Q(t) = \frac{P(t) A_a}{\mu R_t(t)} = \frac{P(t) A_a}{\mu \left(R_0 + R_p(t) + R_c(t) \right)} \tag{88}$$

where $P(t)$ is the total trans-membrane pressure (Pa); A_a is the apparent membrane area (m^2); μ is the liquid viscosity (Pa·s); $R_t(t)$, R_0, $R_p(t)$, and $R_c(t)$, respectively, are the total resistance, the virgin membrane resistance, the pore resistance, and the cake resistance (m^{-1}).

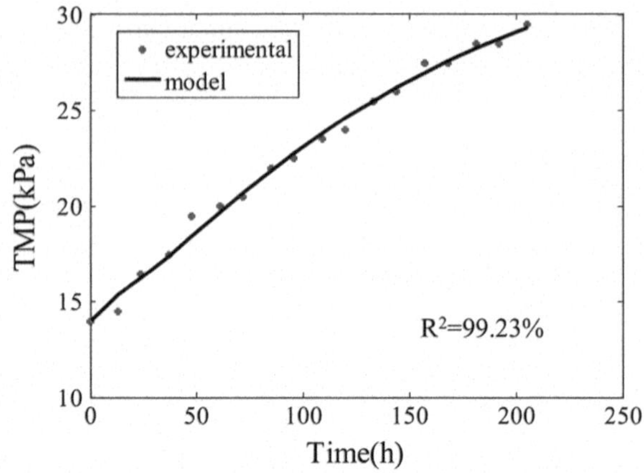

Fig. 5 Model comparison with TMP data obtained with real fluidized bed membrane bioreactor

In the constant flowrate mode, Eq. (88) can be rewritten as

$$\frac{P(t)}{R_t(t)} = \frac{P_m(t)}{R_0 + R_p(t)} = \frac{P_c(t)}{R_c(t)} = \mu J_0 \tag{89}$$

where $P_m(t)$ is the pressure drop across the membrane (originating from the pore blocking/shrinking resistance of the membrane and the inherent membrane resistance), $P_c(t)$ is the pressure drop across cake (Pa) (originating from the resistance of the deposited cake layer), and J_0 is the constant permeate flux (m/s).

As presented in Fig. 6:

$$P(t) = P_m(t) + P_c(t) \tag{90}$$

where P(t) is the total trans-membrane pressure (Pa), $P_m(t)$ is the pressure drops across the membrane (Pa), and $P_c(t)$ is the pressure drop across the cake (Pa).

According to the cake filtration model [15, 17], the $P(t)$ variation with permeate volume ($V(t)$) can be expressed in the following equation:

$$\frac{P(t)}{P_0} = 1 + K_c' J_0 V(t) \tag{91}$$

In cake filtration process, $P_m(t)$ is equal to P_0. Substituting Eq. (91) into Eq. (90), the $P_c(t)$ can be obtained as

$$P_c(t) = P_0 K_c' J_0 V(t) = K_c J_0 V(t) \tag{92}$$

where K_c is cake coefficient (Pa·s/m^2) and $V(t)$ is permeated volume (m^3/m^2).

The $P_m(t)$ can be described by Hagen-Poiseuille law:

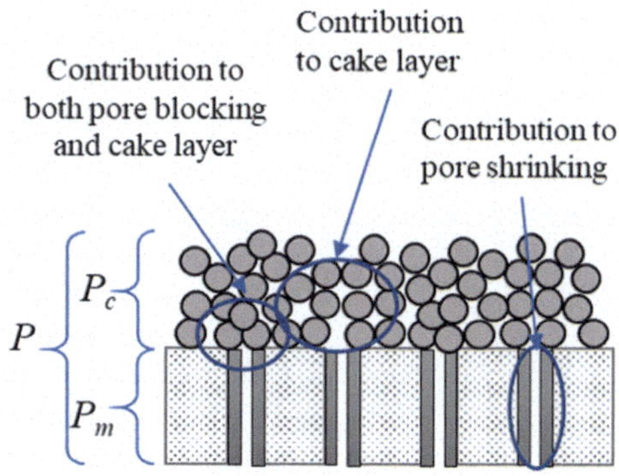

Fig. 6 The schematic of fouling mechanism (**a**) and $P(t)$ variation of the proposed model (**b**)

$$P_m(t) = \frac{32l\mu u}{d^2} \tag{93}$$

where l is the equivalent length of the membrane pore (m), u is flow velocity in membrane pore (equal to Q/A) (m/s), and d is the average pore size of the membrane (m)$\left(d^2 = 4A/n\pi\right)$. In this way, Eq. (93) can be rewritten as

$$P_m(t) = \frac{8\pi nl\mu Q}{A_0^2}\left(\frac{A(t)}{A_0}\right)^{-2} \tag{94}$$

where n is the number of pores, $A_0\left(m^2\right)$ is the total initial pore area at $t = 0$, $A(t)$ (m^2) is total pore area at $t = t$, and $A(t)/A_0$ is regarded as the available membrane area ratio.

According to Eq. (94), P_m(t) is a function of the number of pores (n) and the available membrane area ratio ($A(t)/A_0$). There is some small finite flow through even the "blocked" pores [20]. Therefore, the following assumptions are made in the current modeling:

(1) the number of pores on the membrane is constant in the filtration process;
(2) the pore blocking/shrinking will decrease the pore area.
(3) the decreasing rate of available pore area follow $dA(t)/dV = -Kp\,A(t)$, i.e., the fouling mechanism is intermediate blocking.

The variation of the available membrane area ratio in Eq. (94) can be described by the following equation [1]:

$$\frac{A(t)}{A_0} = e^{-K_p V(t)} \tag{95}$$

where K_p is pore coefficient (m^{-1}) and $V(t)$ is permeated volume (m^3/m^2).

However, adsorption is a dominating mechanism in the early filtration process [21, 22]. Therefore, the available membrane area will reach a constant (limited value) due to the reduction of adsorption and the steady available membrane area ratio in Eq. (95)

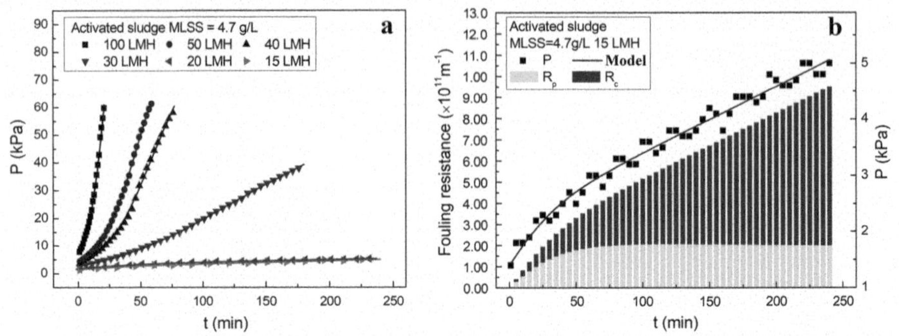

Fig. 7 Model validation by using activated sludge suspension and 0.1μm PAN membranes: **a** $P(t)$ development with filtration time; **b** the variation of $P(t)$, $R_c(t)$, and $R_p(t)$ at 15 LMH

can be corrected as [23]:

$$\frac{A(t)}{A_0} = (1 - K)e^{-K_p J_0 t} + K \tag{96}$$

where K is the steady available membrane area ratio and t is filtration time (s).

The $P(t)$ variation could be obtained by substituting Eqs. (92), (94), and (96) into Eq. (90):

$$P(t) = P_m(t) + P_c(t) = K_m \mu Q \big((1 - K)e^{-K_p J_0 t} + K\big)^{-2} + K_c J_0^2 t \tag{97}$$

where $K_m\left(8\pi nl/A_0^2\right)$ is coefficient (m^{-3}).

The proposed model had good agreements with experimental data under different conditions in a dead-end constant flux microfiltration filtration process ($R^2 = 0.92$–0.99) (Fig. 7).

(4) Yao's model [24]

In constant flowrate mode [17]:

$$\frac{p}{p_0} = \frac{RA_0}{R_0 A} \quad (52)$$

where p_0 is the initial trans-membrane pressure (Pa), R_0 is the initial resistance (m^{-1}), and A_0 is the initial available membrane filtration area (m^2).

For intermediate pore-blocking and cake filtration mechanism [17]:

$$\frac{A}{A_0} = \frac{1}{\exp(K_i J_0 t)} \quad (59)$$

$$\frac{R}{R_0} = \left(1 + K_c J_0 V'\right) \quad (56)$$

where K_i and K_c are constants for intermediate pore-blocking (m^{-1}) and cake filtration laws (s/m^2), respectively.

$\frac{A}{A_0}$ will tend to be constant after it drops to a certain value because of the existence of the cake filtration mode. If assuming the coefficient of available membrane filtration area (ω) represents the effective filtration area would reach a stable value at the timing of t, moreover, considering the effective filtration area (A) is equal to the initially available membrane filtration area (A_0) at time of 0. Thus, the loss of effective filtration area (A/A_0) with filtration time can be expressed as

$$\frac{A}{A_0} = \frac{1-\omega}{\exp(K_i J_0 t)} + \omega \tag{98}$$

Here the constant ω is in the range of 0 to 1 ($0 \leq \omega \leq 1$). Therefore, the value of ω equal to 1 revealing the fouling mechanism is cake filtration. In contrast, the value of ω in the range of 0 to 1 representing the fouling mode is combined intermediate blockage and caking mode.

Then the ideal permeation flux (J') and ideal accumulated volume (V') can be expressed according to Darcy's law, respectively:

$$\frac{J'}{J_0} = \frac{1}{\frac{1-\omega}{\exp(K_i J_0 t)} + \omega} \tag{99}$$

$$V' = \frac{1}{\omega K_i} \ln(\omega \exp(K_i J_0 t) + 1 - \omega) \tag{100}$$

The resistance caused by cake filtration can be written as a function of filtration time by inserting Eq. (100) into Eq. (56):

$$\frac{R}{R_0} = 1 + \frac{K_c J_0}{\omega K_i} \ln(\omega \exp(K_i J_0 t) + 1 - \omega) \tag{101}$$

The variation of trans-membrane pressure as a function of filtration time at constant flux mode can be obtained by inserting Eqs. (98) and (101) into Eq. (52):

$$p = \frac{p_0 \exp(K_i J_0 t)}{\omega \exp(K_i J_0 t) + 1 - \omega} \left(1 + \frac{K_c J_0}{\omega K_i} \ln(\omega \exp(K_i J_0 t) + 1 - \omega)\right) \tag{102}$$

The proposed model had good agreements with experimental data under different conditions in a dead-end constant flux microfiltration filtration process (Fig. 8).

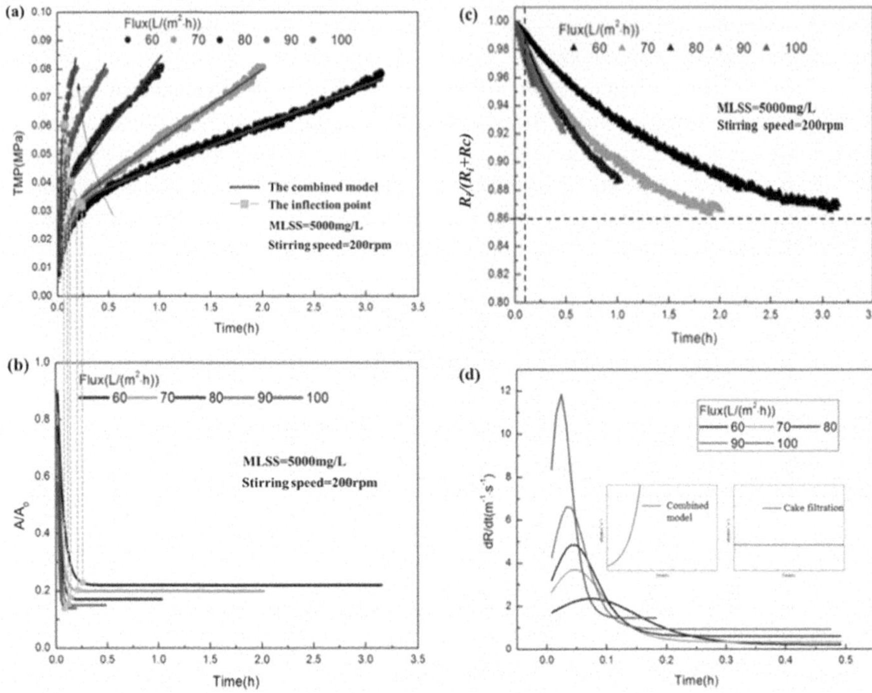

Fig. 8 Evaluation of the proposed model under various flux conditions. **a** variations of instantaneous TMPs; **b** variations of loss of effective filtration area; **c** proportion of resistance of pore blocking in total fouling resistance; **d** variations of fouling rate. Operating conditions: stirring speed = 200 rpm and MLSS = 5000 mg/L

2.6 Combined Tree Mechanism Model

(1) **Hu's Model** [25]

The trans-membrane pressure drops in constant flux mode can be described as [17]

$$\frac{p(t)}{p_0} = \frac{R(t) \cdot A_0}{R_0 \cdot A(t)} \quad (52)$$

where p_0 is the initial trans-membrane pressure drop (Pa), $R(t)$ is the total fouling resistance at filtration time t (m^{-1}), R_0 is the virgin membrane resistance (m^{-1}), $A(t)$ is available membrane area at filtration time t (m^2), and A_0 is the virgin membrane area (m^2), respectively.

The complete pore blocking will reduce the available membrane area, which can be expressed as follows for complete pore-blocking mechanism in the cross-flow mode [26]:

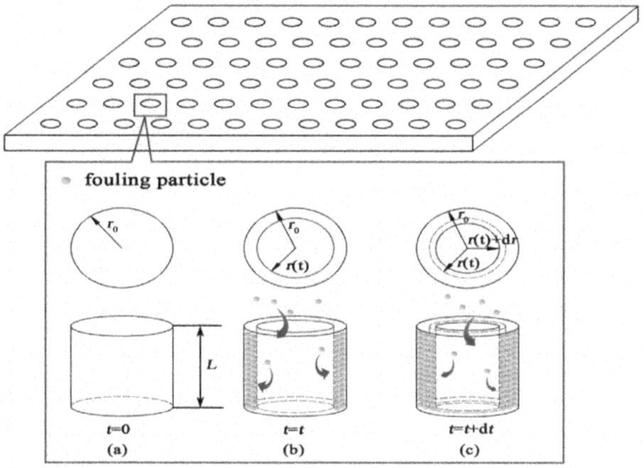

Fig. 9 Schematic diagram for the fouling process in the membrane pores: **a** at the initial fouling process ($t = 0$), **b** at the fouling time t, and **c** at the fouling time $t + \mathrm{d}t$

$$\frac{A(t)}{A_0} = 1 - \frac{K_b}{B}\left(1 - e^{-Bt}\right) \tag{103}$$

where K_b is the complete pore-blocking constant (s^{-1}) and B is the back-transport coefficient of particles from the membrane surface by shear stress (s^{-1}), respectively (Fig. 9).

(i) Effect of complete pore blocking on standard pore blocking

Some particles with diameters less than the pore diameter of the membrane will enter membrane pores, then be adsorbed and deposited on the pore wall, which could reduce the radius of the membrane pores $r(t)$ and the corresponding flow rate related to $r(t)$ can be described by Hagen-Poiseuille equation:

$$Q(t) = \frac{\pi\, r^4(t) p(t)}{8\mu L} N_0 \tag{104}$$

where $Q(t)$ is the volumetric flowrate at filtration time t ($m^3 \cdot s^{-1}$), $r(t)$ is the pore radius at filtration time t(m), $N_0\left(A_0 = N_0 \pi r_0^2\right)$ is the number of the virgin membrane pores in the membrane surface, μ is the viscosity of permeate (Pa·s), and L is the length of pore (m), respectively.

In constant flux mode, after combining Eqs. (52) and (104):

$$\frac{p(t)}{p_0} = \left(\frac{r_0}{r(t)}\right)^4 = \frac{R(t)}{R_0} = \frac{R_s(t) + R_0}{R_0} \tag{105}$$

where r_0 is the initial pore radius on the membrane surface (m) and $R_s(t)$ is the standard pore-blocking resistance (m^{-1}).

In the actual membrane filtration process, membrane pores are often blocked due to the fact that tiny particles can only enter the unblocked pores. And the mass conservation equation within a short period (dt) (Fig. 9) can be written as

$$\pi LN(t)\left[(r(t)-dr)^2-r(t)^2\right]=vJ_0A_0dt \tag{106}$$

$$-L\frac{N(t)}{N_0}\cdot 2r(t)dr=vJ_0r_0^2dt \tag{107}$$

where $N(t)$ is the number of the unblocked membrane pores at filtration time t, v is the particles' volume that remained in the membrane pores per unit filtration volume ($m^3 \cdot m^{-3}$), and $N(t)/N_0$ is equal to $A(t)/A_0$.

After integrating Eq. (107):

$$\left(\frac{r(t)}{r_0}\right)^2=1-\frac{vJ_0}{L(B-K_b)}\left[\ln\left(1-\frac{K_b}{B}\left(1-e^{-Bt}\right)\right)+Bt\right] \tag{108}$$

And the standard pore-blocking resistance can be obtained by substituting Eq. (108) into Eq. (105):

$$R_s(t)=R_0\left[1-\frac{vJ_0}{L(B-K_b)}\left(\ln\left(1-\frac{K_b}{B}\left(1-e^{-Bt}\right)\right)+Bt\right)\right]^{-2}-R_0 \tag{109}$$

(b) Effect of complete pore blocking and shear force on the cake resistance

The cake resistance $R_c(t)$ can be expressed as

$$\frac{R_c(t)}{R_0}=K_cJ_0V(t)' \tag{110}$$

where $R_c(t)$ is the cake resistance (m^{-1}), K_c is the cake filtration constant ($s \cdot m^{-2}$), and $V(t)'$ is the volume filtered through available membrane area at filtration time t ($m^3 \cdot m^{-2}$), respectively.

However, in actual filtration process, foulants usually have wide particle distribution and the fouling mechanism usually is pore blocking and cake filtration. Due to the existence of pore blocking, the available membrane area decreased, and the filtration flux described for the cake formation ($J'(t)$) is changed and it can be expressed as

$$J'(t)=\frac{dV(t)'}{dt}=\frac{J_0}{A/A_0} \tag{111}$$

where $J'(t)$ is the filtration flux related to the available membrane area ($m^3 \cdot m^{-2} \cdot s^{-1}$). And $V(t)'$ can be written as

$$V'(t)=\frac{J_0}{B-K_b}\left[\ln\left(1-\frac{K_b}{B}\left(1-e^{-Bt}\right)\right)+Bt\right] \tag{112}$$

In cross-flow filtration, the shear force can take away some particles from the surface of the cake and the cake resistance at filtration time t can be rewritten as [11]

$$R = R_0 + \alpha\left(k_c V(t)/A - \int_0^t S dt\right) \quad (113)$$

where α is the specific cake resistance (m·kg^{-1}), k_c is the cake filtration constant (kg·m^{-3}), $V(t)$ is the filtration volume at filtration time t (m^3), and S is the erosion rate of cake per unit area at filtration time t (kg·m^{-2}·s^{-1}), respectively.

Thus, the cake resistance can be expressed as

$$R_c(t) = \frac{K_c J_0^2}{B - K_b}\left[\ln\left(1 - \frac{K_b}{B}\left(1 - e^{-Bt}\right)\right) + Bt\right] R_0 - \alpha St \quad (114)$$

(iii) **Establishment of combined model**

The trans-membrane pressure drop can be written as

$$\frac{p(t)}{p_0} = \frac{R_c(t) + R_s(t) + R_0}{R_0} \cdot \frac{A_0}{A(t)} \quad (115)$$

The three-mechanism (complete pore blocking, standard pore blocking, and cake filtration) model in cross-flow constant flux filtration can be obtained by substituting Eqs. (103), (109), and (114) into Eq. (115):

$$\frac{p(t)}{p_0} = \frac{\left[1 - \frac{v J_0}{L(B-K_b)}\left(\ln\left(1 - \frac{K_b}{B}\left(1 - e^{-Bt}\right)\right) + Bt\right)\right]^{-2} + \frac{K_c J_0^2}{B-K_b}\left[\ln\left(1 - \frac{K_b}{B}\left(1 - e^{-Bt}\right)\right) + Bt\right] - \frac{\alpha St}{R_0}}{1 - \frac{K_b}{B}\left(1 - e^{-Bt}\right)} \quad (116)$$

The model is also suitable for the situation when the fouling mechanism is a single mechanism or two mechanism. As the fouling mechanism is converted into two mechanisms, such as complete pore blocking and cake filtration, the corresponding volume of particles remained in membrane pores can be ignored (the parameter v is equal to 0) and Eq. (116) becomes

$$\frac{p(t)}{p_0} = \frac{1 + \frac{K_c J_0^2}{B-K_b}\left[\ln\left(1 - \frac{K_b}{B}\left(1 - e^{-Bt}\right)\right) + Bt\right] - \frac{\alpha St}{R_0}}{1 - \frac{K_b}{B}\left(1 - e^{-Bt}\right)} \quad (117)$$

As the filtration mechanism is converted into a single cake filtration, the complete pore-blocking constant (K_b) and the K_b/B are equal to 0, and the available membrane area ratio (A/A_0) is a constant ($A/A_0 = 1$). In this situation, Eq. (117) can be simplified as

$$\frac{p(t)}{p_0} = 1 + K_c J_0^2 t - \frac{\alpha S t}{R_0} \tag{118}$$

Therefore, the proposed model (Eq. 116) is adaptable, respectively, for a single mechanism (complete pore blocking or standard pore blocking or cake filtration), two mechanism (complete pore blocking and cake filtration) and three mechanism (complete pore blocking, standard pore blocking, and cake filtration) in different filtration processes (Fig. 10).

A good agreement between experimental data and model predictions was observed in Fig. 11.

(2) Gao's model [27]

In constant flow rate mode, P could be divided into the pressure drops across the membrane layer (P_{mpp}) and cake layer (P_c) [28] according to Darcy's equation [1]:

$$J = \frac{P}{\mu R} = \frac{P_{mpp} + P_c}{\mu\left(R_{mpp} + R_c\right)} \tag{119}$$

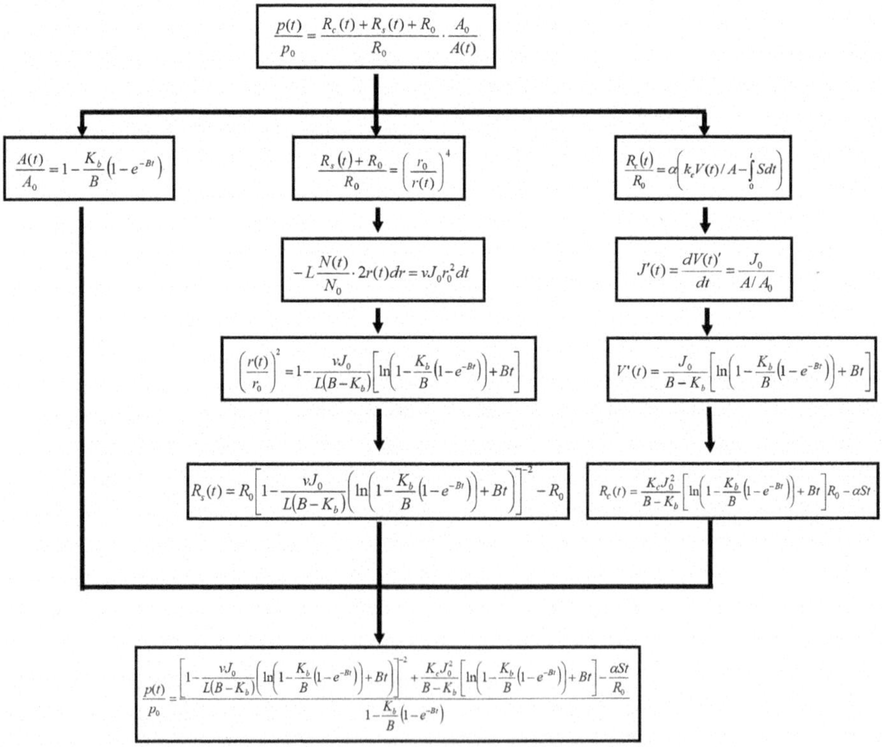

Fig. 10 The flow chart of model derivation

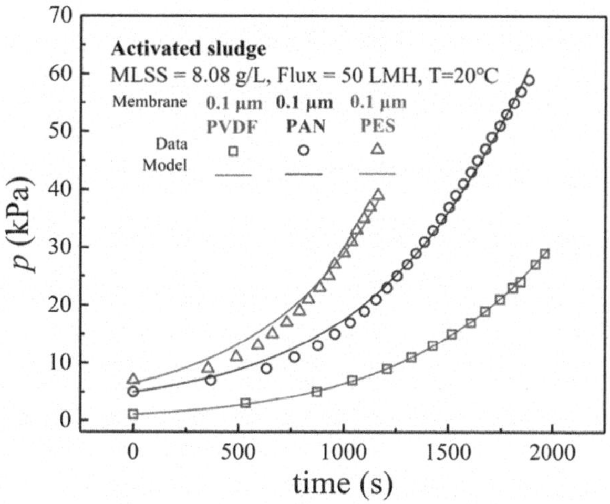

Fig. 11 Experimental data and model predictions for membrane fouling behaviors for 0.1 μ m PAN, 0.1 μ m PVDF, and 0.1 μ m PES membranes at a fixed MLSS concentration of 8.08 g/L and a fixed flux of 50 LMH

$$P = P_{mpp} + P_c \tag{120}$$

where J ($m^3 \cdot m^{-2} \cdot s^{-1}$) is the permeate flux; μ is the solution viscosity (kPa·s); P (kPa), P_{mpp} (kPa), and P_c (kPa) are the total trans-membrane pressure drop, the membrane layer pressure drop, and the cake layer pressure drop, respectively; R (m^{-1}), R_{mpp} (m^{-1}), and R_c (m^{-1}) are the total fouling resistance, the membrane layer resistance, and the cake layer resistance, respectively (Fig. 12).

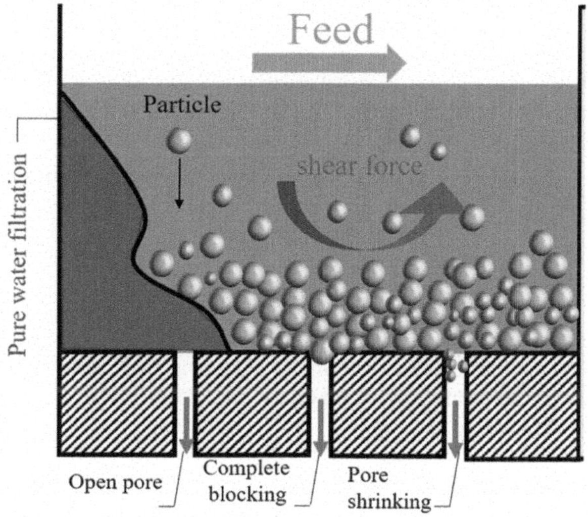

Fig. 12 The schematic of the fouling mechanism

The membrane layer pressure drops (P_{mpp})

The membrane layer pressure drop (P_{mpp}) is caused by pore blocking ($P_{p,b}$)/shrinking ($P_{p,in}$) and the virgin membrane pressure drop (P_m) and can be summed up as

$$P_{mpp} = P_m + P_{p,b} + P_{p,in} \tag{121}$$

where P_m (kPa) is the virgin membrane pressure drop, $P_{p,b}$(kPa) is the pressure drop caused by pore blocking, and $P_{p,in}$(kPa) is the pressure drop caused by pore shrinking.

The pressure drop caused by pore blocking ($P_{p,b}$)

The process of the feed passing through the membrane pores can be regarded as the process of through the capillary in a laminar flow pattern, thus causing the pore blocking, and laminar flow through a straight cylindrical pore can be described using the Hagen-Poiseuille equation [26]:

$$Q = \frac{n\pi r^4 P}{8\mu l} \tag{122}$$

where Q ($m^3 \cdot s^{-1}$) is the volumetric flow rate, n is the number of membrane pores, l (m) is the pore length (i.e., membrane thickness), and r (m) is the pore radius $\left(r^2 = A/n\pi\right)$.

Supposing the fouling mechanism in the membrane pores is only complete pore blocking, the pressure drop caused by pore blocking ($P_{p,b}$) can be written as

$$P_{p,b} = \frac{8n\pi l\mu Q}{A_0^2}\left(\frac{A}{A_0}\right)^{-2} - P_m \tag{123}$$

where A_0 (m^2) is the virgin available membrane area, A(m^2) is the available membrane area at the filtration process (m^2), and A/A_0 is regarded as the available membrane area ratio.

And A/A_0 can be described by complete blocking laws [25]:

$$\frac{A}{A_0} = 1 - \frac{K_b}{B}(1 - \exp(-Bt)) \tag{124}$$

where K_b(s^{-1}) is the complete pore-blocking constant and B(s^{-1}) is the particle resuspension rate, a constant reflecting the frequency at which foulants are removed from the membrane surface by the cross-flow shear force.

Therefore, $P_{p,b}$ can be described as

$$P_{p,b} = K_p\left(\frac{A}{A_0}\right)^{-2} - P_m \tag{125}$$

where $K_p(8\pi nl\mu Q/A_{0^2})$ is coefficient (kPa).

The pressure drop caused by pore shrinking ($P_{p,in}$)

Supposing the only pore shrinking in the membrane pores, the relationship of the pore radius (r) and the pressure drop caused by pore shrinking ($P_{p,in}$) in Eq. (121) can be described by pore shrinking laws [26]:

$$P_{p,in} + P_m = P_m\left(\frac{r_0}{r}\right)^4 \tag{126}$$

where r_0(m) is the pore radius of the virgin membrane and r(m) is the pore radius in the filtration process.

Following Hermia, a solids mass balance yields

$$n\pi\left(r_0^2 - r^2\right)l = CV \tag{127}$$

where C is the volume of particles deposited per unit of filtrate and $V(\text{m}^3)$ is the permeate volume that has passed through the membrane.

Rearranging Eqs. (126) and (127):

$$\left(\frac{r}{r_0}\right)^4 = \left(1 - \frac{CV}{n\pi l r_0^2}\right)^2 \tag{128}$$

Therefore, $P_{p,in}$ can be described as

$$P_{p,in} = P_m\left((1 - K_{in}A_0Jt)^{-2} - 1\right) \tag{129}$$

where the K_{in} (m^{-3}) is the pore shrinking constant and defined as

$$K_{in} = \frac{C}{n\pi l r_0^2} \tag{130}$$

The cake layer pressure drops (P_c)

As the filtration mechanism is converting into single cake filtration, an erosion term is added to account for the removal of foulant layers due to cross flow, the variation of the cake layer pressure drop (P_c) can be expressed [25, 26] as

$$\frac{P_c}{P_m} = 1 + K_c'J^2t - \frac{\alpha St}{R_m} \tag{131}$$

where α ($\text{m}\cdot\text{kg}^{-1}$) is the specific cake resistance, S ($\text{kg}\cdot\text{m}^{-2}\cdot\text{s}^{-1}$) is the erosion rate of the cake per unit area at filtration time t (s), and $K_c'\left(\text{s}\cdot\text{m}^{-2}\right)$ is the cake filtration constant for cross-flow filtration.

In the cake filtration process, P_c can be obtained as

$$P_c = P_m\left(K_c' J^2 - \frac{\alpha S}{R_m}\right)t = K_c J^2 t \tag{132}$$

where $K_c(\text{s} \cdot \text{m}^{-2})$ is the modified cake filtration constant for cross-flow filtration and is defined as

$$K_c = P_m K_c' - \frac{\alpha S}{J^2 R_m} \tag{133}$$

Establishment of a combined model

Finally, the variation of total P could be written as

$$P = K_p\left(\frac{A}{A_0}\right)^{-2} + P_m((1 - K_{in} A_0 J t)^{-2} - 1) + K_c J^2 t \tag{134}$$

The derivation above is very rough and the detailed model establishment is shown as follows but with a flow chart of model derivation shown firstly (Fig. 13).

A good agreement between experimental data and model predictions was observed ($R^2 > 0.992$ for SA, $R^2 > 0.874$ for HA and $R^2 > 0.985$ for SA) (Fig. 14).

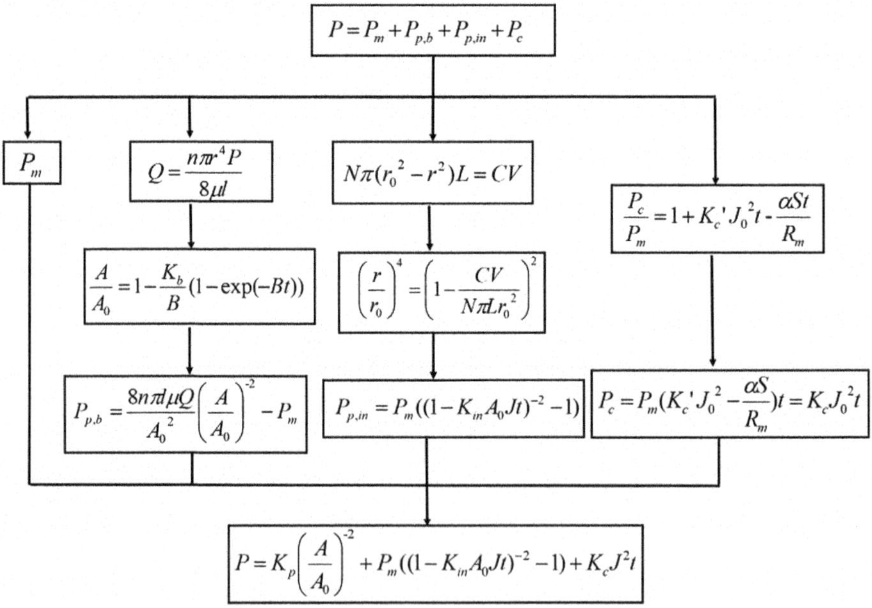

Fig. 13 The flow chart of model derivation

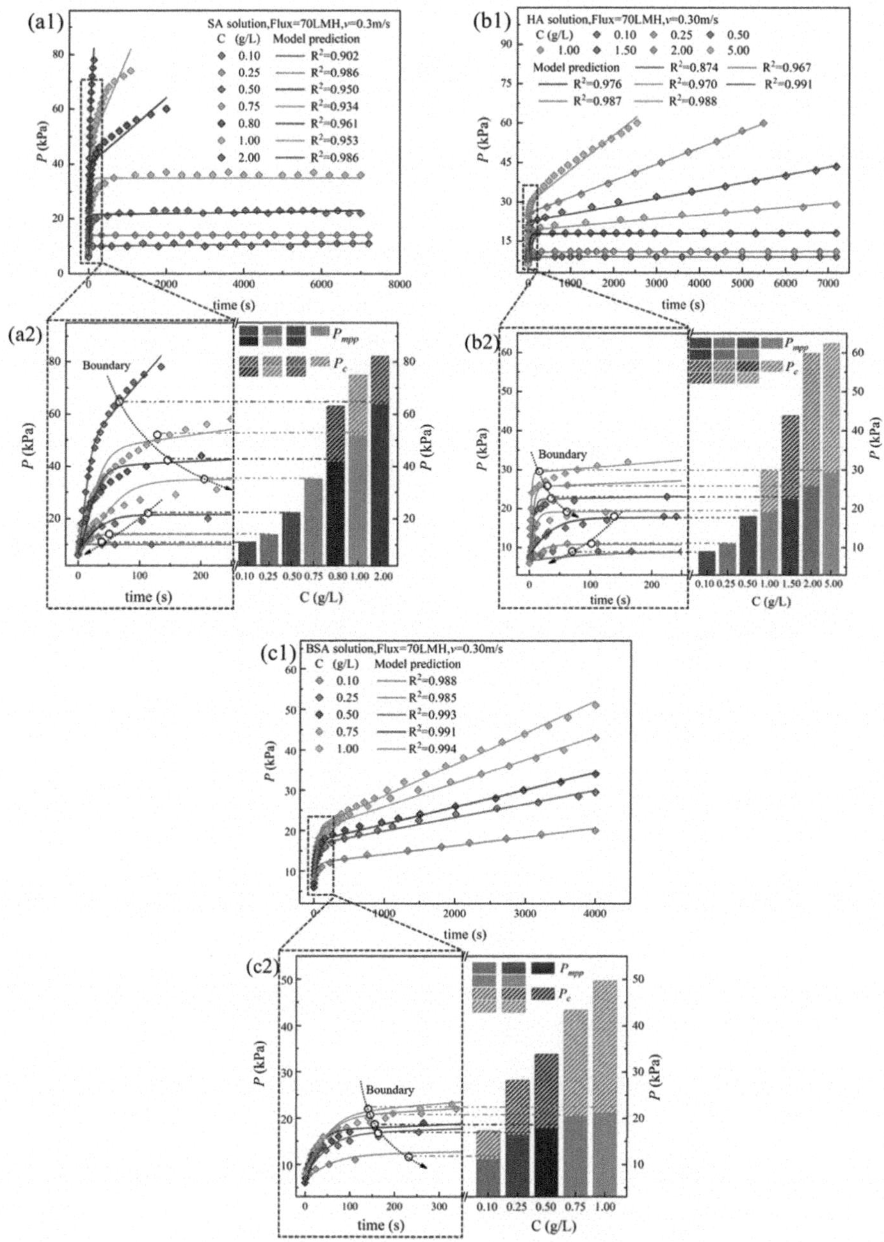

Fig. 14 The variation of the total pressure drop (P) ((a1), (b1), and (c1)); the values of the membrane layer pressure drop (P_{mpp}) and cake layer pressure drop (P_c) at the end of the filtration process ((a2), (b2), and (c2)) of 0.1 μm PAN membrane, respectively, fouled with SA, HA, and BSA at different concentrations

2.7 Other Model

(1) **Semi-empirical model** [29]

According to Darcy's law:

$$J = \frac{1}{A_c}\frac{dV}{dt} = \frac{\Delta P_L}{\mu(R_c + R_m)} \tag{135}$$

where J is the flux (m/s), P_L id trans-membrane pressure (Pa), R_m is the membrane resistance (m^{-1}), and R_c is the cake resistance (m^{-1}).

The cake resistance (R_c) also can be expressed as

$$R_c = \alpha\frac{C_s V}{A_c} \tag{136}$$

where α is the specific filter cake resistance (m/kg) and C_s is the solid volume concentration (kg/m^3), V is cumulative filtrate volume (m^3) and A_c is cake area (m^2).

After substituting it into Eq. (135), we obtain

$$\frac{t}{V} = \frac{\mu C_s \alpha}{2A_c\Delta P_L}V + \frac{\mu R_m}{A_c\Delta P_L} \tag{137}$$

If the flowrate is constant, from Eq. (137), we obtain

$$\Delta \mathrm{P_L} = \alpha\mu C_s\frac{Q^2}{A_c^2}t + \mu R_m\frac{Q}{A_c} \tag{138}$$

ΔP_L can be expressed as

$$\Delta \mathrm{P_L} = \Delta \mathrm{P_c} + \Delta \mathrm{P_m} \tag{139}$$

where ΔP_c is the pressure across the cake (Pa) and ΔP_m is the pressure across the membrane (Pa).

For compressible materials, the specific filter cake resistance can be expressed as

$$\alpha = \alpha_0 \Delta P_L^n \tag{140}$$

where α_0 is specific cake resistance at zero pressure drop (m/kg).

$$(\Delta P_c)^{1-n} = \alpha_0(1-n)\mu c\frac{Q^2}{A^2}t \tag{141}$$

The pressure drop across the membrane is constant and is expressed by the following formula:

$$\Delta P_m = \frac{\mu R_m Q}{A_c} \tag{142}$$

The transient effect can be approximated as a single process and modeled according to the following equation:

$$\alpha(\Delta P_L, t) = \alpha_0\left(1 - e^{-t/\tau}\right)\Delta P_L^n \tag{143}$$

where τ is the average consolidation time constant.

After substituting Eq. (141) into Eq. (138), we obtain

$$\Delta P_L = \alpha_0\left(1 - e^{-t/\tau}\right)\Delta P_L^n \mu C_s \frac{Q^2}{A_c^2} t + \mu R_m \frac{Q}{A_c} + \Delta P_0 \tag{144}$$

When a shear force is applied to the filter cake (assuming that the applied shear force does not remove the filter cake), the pressure required to compress the material to a given solids fraction is smaller. The overall compression coefficient (n) is expected as

$$n = \gamma_1 G + \gamma_2 \tag{145}$$

where γ_1 and γ_2 can be determined empirically from the experimental data of TMP rise at different shear rates.

$$\alpha_L = a_0(1 - e^{-\frac{t}{\tau}})\Delta P_L^n(\gamma_1 G + \gamma_2) \tag{146}$$

The constant flux is modified to a general equation containing the shear term.

$$\Delta P_L = a_0(1 - e^{\frac{t}{\tau}})\Delta P_L^{(\gamma_1 G + \gamma_2)} \mu C_s \frac{Q^2}{A_c^2} t + \Delta R_m + \Delta P_0 \tag{147}$$

For the coni-cylindrical device as presented in Fig. 15

The filtration cell containing the coni-cylindrical Couette was pressurized allowing for simultaneous shear and filtration (shown in Fig. 15).

$$\mathrm{G} = \frac{\kappa_R \Omega}{1 - \kappa_R} \tag{148}$$

where $\kappa_R = r_a/r_b$, r_a is the inner radius and r_b is the outer radius. The shear environment adjacent to the membrane bottom is essentially a cone and plate arrangement. A simple relationship that gives the shear rate in such an environment is given by Mooney and Ewart:

$$\mathrm{G} = \frac{\Omega}{\theta} \tag{149}$$

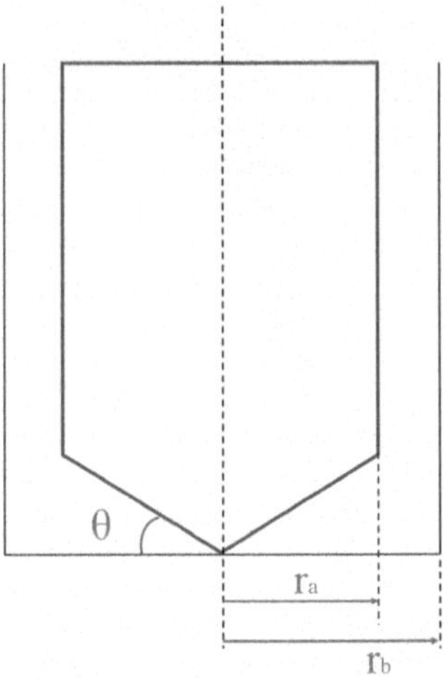

Fig. 15 Schematic of coni-cylindrical Couette with key dimensions of the rotating inner part, outer cell, and cone angle which govern the shear environment

where Ω is the rotational Couette speed and θ is the cone angle. The vessel was designed with a modest κ_R of 0.9. At operational shear rates below 12 s^{-1} Taylor vortices are not expected to be significant and a range of low-shear environments that are reasonably uniform can be generated.

A good agreement between experimental data and model predictions was observed (Fig. 16).

(2) **Modified classic pore blocking law via probability** [30]

Classic pore blocking law, which Hermia proposed, tells the truth that particles have already deposited on the membrane surface to form a cake layer attached to the membrane pore wall to shrink pores or block the membrane pores. However, a

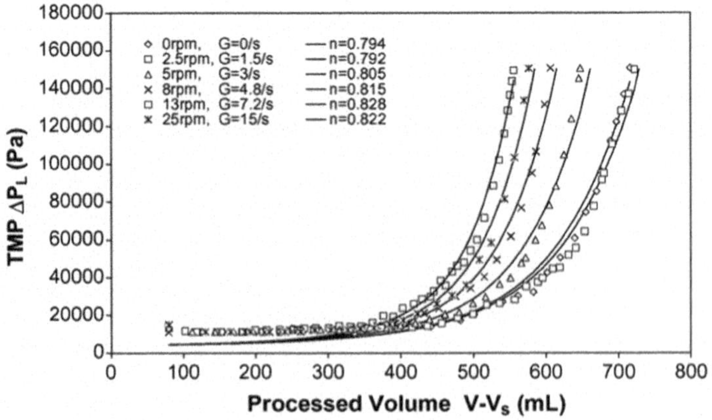

Fig. 16 Equation (147) fitted to the experimental data extracting compressibility values (n) for each shear rate with a common time constant of $\tau = 77,600$ s

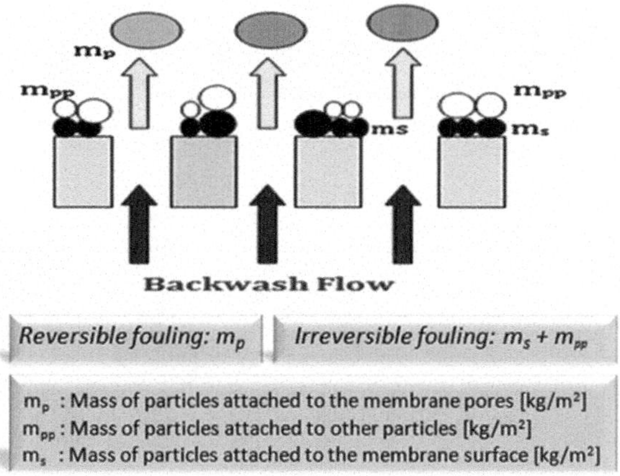

Fig. 17 Irreversible cake fouling after backwash

particle that collides with another particle or the membrane surface can attach with them or leave it away. In 2013, Abdelrasoul modified the classic pore-blocking law using probability [30].

In Abdelrasoul's model, one should emphasize that reversible fouling refers to the particles that cause pore blocking (Fig. 17). Owing to the homogeneous surface chemical properties, the probability of a particle attached to the membrane surface is the same everywhere $\left(\alpha_{pm}\right)$ and particle-to-particle attachment $\left(\alpha_{pp}\right)$. Meanwhile, the objective particle is latex which is assumed to be a rigidity ball. The sum of probabilities of attachment is always less than 1 because of that not all collision results in attachment. This model can be established by employing these assumptions in the following two significant cases.

Case 1. The fouling particle is more significant than the membrane pore ($a > D_m/2$).

It is generally accepted that the sum of probabilities of collision is equal to 1:

$$\beta_s + \beta_p + \beta_{pp} = 1 \tag{150}$$

where β_p is the collision probability between a particle and the membrane surface (Fig. 18, particle 1),β_p is the collision probability between a particle and a membrane pore (particle 2), and β_{pp} is the collision probability between a particle and another particle (particle 3).

β_{pp} is proportional to the membrane area ratio, which other particles have covered. Hence, an uncovered area ratio (A_t) equation can be established as follows:

$$\mathrm{d}A_t = -4\alpha_{pp}\sigma A_t \mathrm{d}m_t \tag{151}$$

where σ is the projected area of a unit mass of particles on the membrane surface and mt is the total mass of particles retained which is equal to $C_f \cdot V_s$ (C_f is the feed concentration, V_s is the permeate volume).

Giving the initial condition of $A_t = 1$ when $m_t = 0$, and after integrating Eq. (151):

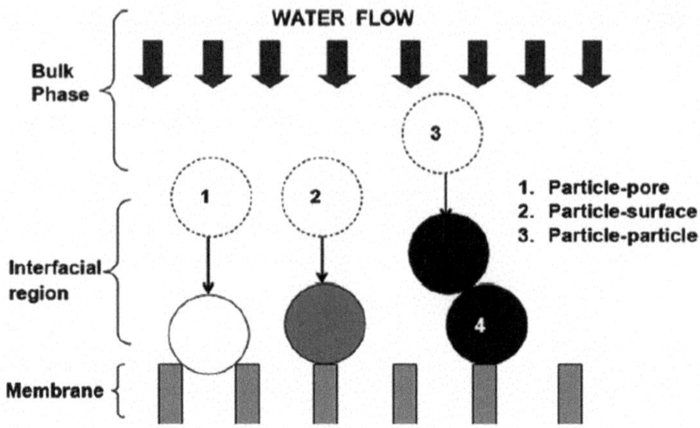

Fig. 18 A diagram of UF membrane fouling by particles larger than the membrane pore size

$$A_t = \beta_s + \beta_{\mathrm{p}} = \exp(-4\alpha_{pp}\sigma m_t) \tag{152}$$

$$\beta_{pp} = 1 - \exp(-4\alpha_{pp}\sigma m_t) \tag{153}$$

By introducing parameter ε_p (the membrane surface porosity), β_s and β_p can be expressed, respectively, as

$$\beta_s = \varepsilon_p \exp(-4\alpha_{pp}\sigma m_t) \tag{154}$$

$$\beta_p = (1 - \varepsilon_p)\exp(-4\alpha_{pp}\sigma m_t) \tag{155}$$

Now, all collision probabilities are expressed via attachment probabilities. Consequently, the procedure of particle depositing to block membrane pores can be assumed that a collision between a particle and a membrane pore happens first, followed by attachment but not leaving. Therefore, the expression of the mass of particles blocked pores (m_p) can be written as

$$\mathrm{d}m_p = \alpha_{pm}\beta_p \mathrm{d}m_t \tag{156}$$

$$m_{pL} = \frac{\alpha_{pm}\varepsilon_p}{4\alpha_{pp}\sigma_L}\left[1 - \exp(-4\alpha_{pp}\sigma_L m_t)\right] \tag{157}$$

The mass of particles deposited on the membrane (m_s) or attached to other particles (m_{pp}) can also be obtained as follows:

$$\mathrm{d}m_s = \alpha_{pm}\beta_s \mathrm{d}m_t \tag{158}$$

$$m_{sL} = \frac{\alpha_{pm}(1-\varepsilon_p)}{4\alpha_{pp}\sigma_L}\left[1-\exp(-4\alpha_{pp}\sigma_L m_t)\right] \quad (159)$$

$$\mathrm{d}m_{pp} = \alpha_{pp}\beta_{pp}\mathrm{d}m_t \quad (160)$$

$$m_{ppL} = \alpha_{pp}m_t - \frac{\alpha_{pp}}{4\alpha_{pp}\sigma_L}\left[1-\exp(-4\alpha_{pp}\sigma_L m_t)\right] \quad (161)$$

The subscript L means the parameter for the large particles. The irreversible fouling particle mass can be calculated as follows:

$$m_{cIRRL} = m_{sL} + m_{ppL} = \alpha_{pp}m_t + \frac{\alpha_{pm}(1-\varepsilon_p)-\alpha_{pp}}{4\alpha_{pp}\sigma_L}\left[1-\exp(-4\alpha_{pp}\sigma_L m_t)\right] \quad (162)$$

Here, all kinds of mass of the particle can be calculated in case 1.

Case 2. Fouling particle is smaller than membrane pore ($a < D_m/2$).

In this case, a particle can enter membrane pores attaching to the pore walls ($a < D_m/6$), or block membrane pores via bridging with other particles ($D_m/6 < a < D_m/2$), or aggregate with each other to form a cake layer, which, respectively, results in pore constriction, pore blocking, and cake formation (Fig. 19).

When the attractive force between particles and membrane pore walls is larger than the drag force caused by permeation, particles can attach to the membrane pore walls. Therefore, the mass of particles attaching to the membrane pore walls (m_w) can be calculated as follows:

$$m_w = \frac{N_m}{\sigma_{xS}\alpha_{pm}B}\left\{\pi D_m L_m B\alpha_{pm} - \ln\left[1+\exp\left(\pi D_m L_m B\alpha_{pm} - \frac{\sigma_{xS}\alpha_{pm}B}{N_m}C_{in}V_s\right)\right]\right\} \quad (163)$$

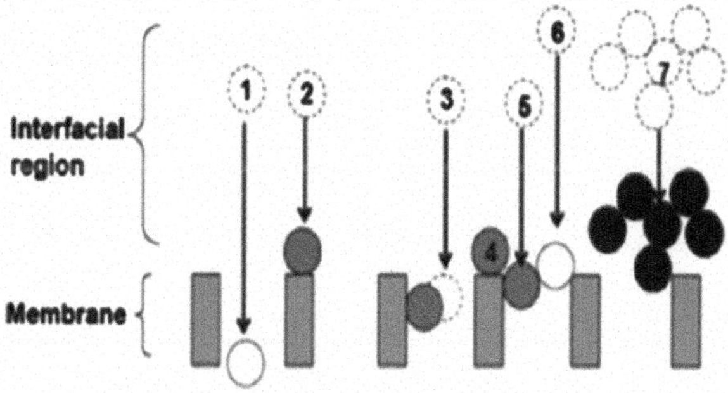

Fig. 19 A diagram of UF membrane fouling by particles smaller than the membrane pore size

where N_m is the pore density in the membrane, L_m is the length of the membrane pore, C_{in} is the mass concentration of particles at the inlet of the membrane pores, and B is the mass transfer coefficient which can be expressed as follows:

$$B = \frac{2D}{D_m Q_1 \ln(D_m/2a)} \tag{164}$$

where D is the diffusion coefficient and Q_1 is the permeate flow rate in a single membrane pore.

Similar to the derivation of m_s, m_p, and m_{pp} for large particles, their corresponding value for small particles ($D_m/6 < a < D_m/2$) can be written as

$$m_{sS} = \frac{\alpha_{pm}(1-\varepsilon_p)}{4\alpha_{pp}\sigma_S}\left[1 - \exp(-4\alpha_{pp}\sigma_S m_t)\right] \tag{165}$$

$$m_{pS} = \frac{\alpha_{pm}\varepsilon_p}{4\alpha_{pp}\sigma_S}\left[1 - \exp(-4\alpha_{pp}\sigma_S m_t)\right] \tag{166}$$

$$m_{ppS} = \alpha_{pp} m_t - \frac{\alpha_{pp}}{4\alpha_{pp}\sigma_S}\left[1 - \exp(-4\alpha_{pp}\sigma_S m_t)\right] \tag{167}$$

Combining case 1 and case 2, the mass of the cake layer can be calculated as follows:

$$m_c = m_t - m_{pL} - m_{pS} - m_w \tag{168}$$

Moreover, the classic pore-blocking law can now be modified.

For complete pore blocking (n = 2), one has

$$\frac{dP}{dm_p} = \sigma P^2 \tag{169}$$

By integrating for case 1 or 2, respectively, to gain the following results:

$$P_{pL} = \frac{1}{1 - \frac{\alpha_{pm}\varepsilon_p}{4\alpha_{pp}}[1 - \exp(-4\alpha_{pp}\sigma_L C_f V_s)]} \tag{170}$$

$$P_{pS} = \frac{1}{1 - \alpha_{pm}\varepsilon_p[1 - \exp(-\sigma_S C_f V_s)]} \tag{171}$$

For standard pore blocking (n = 1.5), one has

$$\frac{dP}{dm_w} = \left(\frac{2}{L_m \rho}\right) P^{1.5} \tag{172}$$

After integrating to obtain

$$P_w = \left\{1 - \frac{N_m}{\sigma_{xS}\rho\alpha_{pm}BL_m}\left\{\pi D_m L_m B\alpha_{pm} - \ln\left[1 + \exp\left(\pi D_m L_m B\alpha_{pm} - \frac{\sigma_{xS}\alpha_{pm}B}{N_m}C_{in}V_s\right)\right]\right\}\right\}^{-2} \tag{173}$$

For cake filtration (n = 0), one has

$$\frac{\mathrm{d}P}{\mathrm{d}m_c} = \frac{R_c}{R_m} \tag{174}$$

After integrating to gain.

$$\begin{aligned} P_c = 1 + \frac{R_c}{R_m}\Bigg\{ & C_f V_s - \frac{\alpha_{pm}\varepsilon_p}{4\alpha_{pp}\sigma_L}[1 - \exp(-4\alpha_{pp}\sigma_L m_t)] \\ & - \frac{\alpha_{pm}\varepsilon_p}{\sigma_S}[1 - \exp(-\sigma_S m_t)] - \frac{N_m}{\sigma_{xS}\alpha_{pm}B}\{\pi D_m L_m B\alpha_{pm} \\ & - \ln\left[1 + \exp\left(\pi D_m L_m B\alpha_{pm} - \frac{\sigma_{xS}\alpha_{pm}B}{N_m}C_{in}V_s\right)\right]\}\Bigg\}. \end{aligned} \tag{175}$$

Therefore, the total trans-membrane pressure can be calculated via the summation of Eqs. (170), (171), (173), and (175).

References

1. J. Hermia, Constant pressure blocking filtration laws—application to power-law non-Newtonian fluids. Trans. Inst. Chem. Eng. **60**, 183–187 (1982)
2. H.P. Grace, Structure and performance of filter media. II. Performance of filter media in liquid service. AIChE J. **2**, 316–336 (1956)
3. M. Shirato, T. Aragaki, R. Mori, K. Sawamoto, Predictions of constant pressure and constant rate filtrations based upon an approximate correction for side wall friction in compression permeability cell data. J. Chem. Eng. Japan **1**, 86–90 (1968)
4. M. Hlavacek, F. Bouchet, Constant flowrate blocking laws and an example of their application to dead-end microfiltration of protein solutuions. J. Membr. Sci. **82**, 285–295 (1993)
5. D.R. Sperry, Note and correspondence: A study of the fundamental laws of filtration using plant-scale equipment. Ind. Eng. Chem. **13**, 1163–1164 (1921)
6. E. Iritani, Y. Mukai, E. Hagihara, Measurements and evaluation of concentration distributions in filter cake formed in dead-end ultrafiltration of protein solutions. Chem. Eng. Sci. **57**, 53–62 (2002)
7. S. Chellam, W.D. Xu, Blocking laws analysis of dead-end constant flux microfiltration of compressible cakes. J. Colloid Interface Sci. **301**, 248–257 (2006)
8. J.A. Suarez, J.M. Veza, Dead-end microfiltration as advanced treatment for wastewater. Desalination **127**, 47–58 (2000)
9. W. Kozicki, C.H. Chou, C. Tiu, Non-Newtonian flow in ducts of arbitrary cross section. J. Chem. Eng. Sci. **21**, 665–679 (1965)
10. S. Kasemset, Z.W. He, D.J. Miller, B.D. Freeman, M.M. Sharma, Effect of polydopamine deposition conditions on polysulfone ultrafiltration membrane properties and threshold flux during oil/water emulsion filtration. Polymer **97**, 247–257 (2016)

11. R.W. Field, D. Wu, J.A. Howell, B.B. Gupta, Critical flux concept for microfiltration fouling. J. Membr. Sci. **100**, 259–272 (1995)
12. R.W. Field, J.J. Wu, Modelling of permeability loss in membrane filtration: Re-examination of fundamental fouling equations and their link to critical flux. Desalination **283**, 68–74 (2011)
13. P. Emami, F. Fallahianbijan, E. Dinse, S.P. Motevalian, B.C. Conde, K. Reilly, A.L. Zydney, Modified intermediate pore blockage model describes fouling behavior during sterile filtration of glycoconjugate vaccines. J. Membr. Sci. **613** (2020)
14. W.B. Richard, *Membrane Technology and Applications* (John Wiley & Sons Ltd, 2012)
15. E. Iritam, N. Katagiri, T. Takenaka, Y. Yamashita, Membrane pore blocking during cake formation in constant pressure and constant flux dead-end microfiltration of very dilute colloids. Chem. Eng. Sci. **122**, 465–473 (2015)
16. B.F. Ruth, Studies in filtration III. Derivation of general filtration equations. Ind. Eng. Chem. **27**, 708–723 (1935)
17. G. Bolton, D. LaCasse, R. Kuriyel, Combined models of membrane fouling: Development and application to microfiltration and ultrafiltration of biological fluids. J. Membr. Sci. **277**, 75–84 (2006)
18. A. Charfi, M. Aslam, J. Kim, Modelling approach to better control biofouling in fluidized bed membrane bioreactor for wastewater treatment. Chemosphere **191**, 136–144 (2018)
19. D. Xinpei, Z. Wang, L. Chao, A combined model for describing thetrans-membrane pressure (P) variation in constant flux dead-end microfiltration (MF) process. J. Environ. Chem. Eng. **10**, 107076 (2022)
20. C. Ho, A.L. Zydney, A combined pore blockage and cake filtration model for protein fouling during microfiltration. J. Colloid Interface Sci. **232**, 389–399 (2000)
21. S. Wu, X. Hua, B. Ma, H. Fan, R. Miao, M. Ulbricht, C. Hu, J. Qu, Three-dimensional analysis of the natural organic matter distribution in cake layer to precisely reveal ultrafiltration membrane Foulingn mechanism. Environ. Sci. Technol. **55**, 5442–5452 (2021)
22. H.T. Lay, R.J.E. Yeow, Y. Ma, A.L. Zydney, R. Wang, J.W. Chew, Internal membrane fouling by proteins during microfiltration. J. Membr. Sci. **637**, 119589 (2021)
23. L. Hou, Z. Wang, P. Song, A precise combined complete blocking and cake filtration model for describing the flux variation in membrane filtration process with BSA solution. J. Membr. Sci. **542**, 186–194 (2017)
24. W. Yao, L. Hou, F. Wang, Z. Wang, H. Zhang, Dual-objective for the mechanism of membrane fouling in the early stage of filtration and determination of cleaning frequency: A novel combined model. J. Membr. Sci. **647**, 120315 (2022)
25. G. Hu, X.S. Liu, Z. Wang, X.P. Du, X. Wang, Comparison of fouling behaviors between activated sludge suspension in MBR and EPS model solutions: A new combined model. J. Membr. Sci. **621** (2021)
26. A.Y. Kirschner, Y.H. Cheng, D.R. Paul, R.W. Field, B.D. Freeman, Fouling mechanisms in constant flux crossflow ultrafiltration. J. Membr. Sci. **574**, 65–75 (2019)
27. S. Gao, C. Zhong, Z. Wang, A model describing the pressure drops across the membrane and cake layers for constant flux cross-flow microfiltration process. J. Environ. Chem. Eng. **10**, 107984 (2022)
28. Z. Zhu, Z. Wang, H. Wang, Y. Kong, K. Gao, Y. Li, Cake properties as a function of time and location in microfiltration of activated sludge suspension from membrane bioreactors (MBRs). Chem. Eng. J. **302**, 97–110 (2016)
29. P. Kovalsky, G. Bushell, T.D. Waite, Prediction of transmembrane pressure build-up in constant flux microfiltration of compressible materials in the absence and presence of shear. J. Membr. Sci. **344**, 204–210 (2009)
30. A. Abdelrasoul, H. Doan, A. Lohi, A mechanistic model for ultrafiltration membrane fouling by latex. J. Membr. Sci. **433**, 88–99 (2013)

Water Rinsing

Nomenclature

J_0	Constant pure water flux of the virgin membrane or rinsed membrane (m/s)
$J(t)$	The instantaneous membrane flux (m/s)
k	Rinsing rate constant (s^{-1})
k_1	Rinsing rate constant for the first part of the rinsing (s^{-1})
k_1	Removal rate constant of loose cake (s^{-1})
k_2	Rinsing rate constant for the second part of the rinsing (s^{-1})
k_2	Removal rate constant of compact cake (s^{-1})
k_3	Re-deposition rate constant of foulants (s^{-1})
k_1	Rate constant of swelling (gel) (s^{-1})
k_2	Rate constant of dissolution (cake) (s^{-1})
k_3	Rate constant of re-deposition (dissolved cake) (s^{-1})
k_4	Rate constant of dissolution (swelled gel) (s^{-1})
k_5	Rate constant of re-deposition (dissolved gel) (s^{-1})
L	Cake thickness (m)
L_0	Initial cake thickness (m)
N	Number of foulant particles on the membrane surface
N_0	Initial number of foulant particle on the membrane surface
p	Instantaneous trans-membrane pressure of the rinsed membrane (Pa)
p_m	Trans-membrane pressure of the virgin membrane (Pa)
r	Specific cake resistance (m^{-1}/m)
R	The current resistance during fouling (m^{-1})
$R_{compact}(0)$	Fouling resistance of compact cake at rinsing time 0 s (m^{-1})
$R_{compact}(t)$	Instantaneous fouling resistance of compact cake (m^{-1})
$R_f\ (0)$	Total fouling resistance at rinsing time of 0 s (m^{-1})
$R_f\ (t)$	Total fouling resistance at rinsing time of t (m^{-1})
$R_{loose}(0)$	Fouling resistance of loose cake at rinsing time 0 s (m^{-1})
$R_{loose}(t)$	Instantaneous fouling resistance of loose cake (m^{-1})

Z. Wang (ed.), *Modelling of the Microfiltration Process*,
https://doi.org/10.1007/978-981-99-0471-6_4

R_m	Cirgin membrane resistance (m^{-1})
R_r	The membrane hydraulic resistance after rinsing (m^{-1})
$R_A(t)$	The fouling resistances of virgin foulants (gel and cake) (m^{-1})
$R_B(t)$	The fouling resistances of the swelled gel (m^{-1})
$R_c(t)$	The fouling resistances of the direct dissolved cake (m^{-1})
R_{cake}	Cake resistance (m^{V1})
$R_D(t)$	The fouling resistances of the dissolved gel (D) (m^{-1})
R_{gel}	Gel resistance (m^{-1})
R_{pore}	Pore resistance (m^{-1})
$R_f(0)$	The fouling resistance t=0 moment (m^{-1})
Rtf	Total fouling resistance (m^{-1})
R_i	Minimum fouling resistance (m^{-1})
R	Instantaneous fouling resistance in the rinsing process (m^{-1})
R_{loose}	Instantaneous fouling resistance of the rinsed loose cake (m^{-1})
R_m	Virgin membrane resistance (m^{-1})
S	Membrane surface (m^2)
u	Cross-flow velocity in the rinsing process (m/s)
u_0	Minimum cross-flow velocity required for foulants removal in the rinsing process (m/s)
u_f	Filtration cross-flow velocity (m/s)
V	Cake volume (m^3)
t	Rinsing time (s)

Greek Symbols

μ	Liquid viscosity (Pa s)
σ	Relative standard error
α	Significant level
ε_c	Cake porosity
δ_1	The characteristic parameter of interaction energy for specific sludge suspension (kT)
δ_2	The parameter of the detachment rate (s^{-1})

1 Introduction

Membrane cleaning is an inescapable and essential step in maintaining membrane filtration processes [1]. Many physical and chemical cleaning methods have been employed to remove the deposited layers on the membrane surface and foulants in the membrane's pores [2] and alleviate the membrane fouling and prolong the service life of the membrane (Fig. 1).

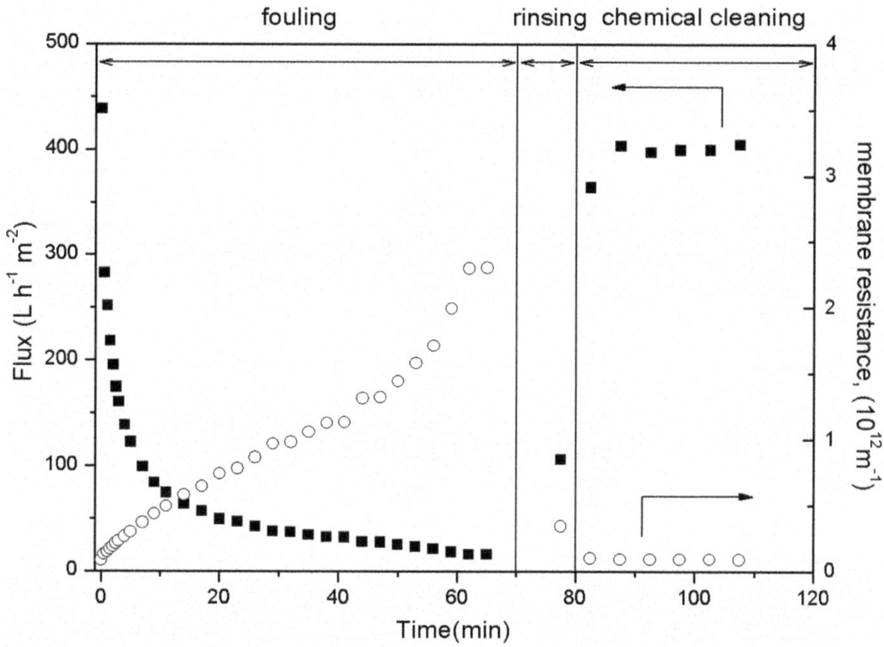

Fig. 1 Typical flux decline and fouling resistance increase in the fouling and cleaning process

Rinsing is one of the most frequently used cleaning operations in the MBR [3] due to it has less damage to the membrane structure and can effectively remove the reversible foulants from the membrane [4], which is controlled by the synergetic cleaning effect [swelling, dissolution, and hydrodynamic effect (shear force)] rather than chemical reaction (Fig. 2).

As an essential physical cleaning method, water rinsing mainly uses the shear force generated by water flow to remove foulants on the membrane surface [5]. Therefore, the efficiency of rinsing was influenced by temperature, trans-membrane pressure, stirring speed (flow velocity), rinsing times, detergent volume, filtration condition, membrane type and feed property, and so on (Fig. 3) [6–8]. (1) Prolonging the flushing time, increasing the flushing rate, and heating can reduce the percentage of non-removable resistance to total fouling resistance (*RRNR*) in the water-washing process. For example, Nakanishi et al. [6] increased the rinsing duration for the CA ultrafiltration membrane fouled by skimmed milk from 10 to 60 min, and the RRNR decreased by 3% while the washing velocity increased from 1.5 m/s to 5.0 m/s, the corresponding RRNR is reduced by 22.2%. Kong et al. [9] raised from 20 °C to 50 °C, and the RRNR decreased by 8% for the PAN membrane fouled by sodium alginate solution in the rinsing process. (2) Reducing the rinsing pressure can reduce RRNR. Nakanishi et al. [6] indicated that the trans-membrane pressure in the water rinsing process was reduced from 3.35 bar to 1.03 bar for the CA ultrafiltration membrane fouled by skimmed milk, and the corresponding RRNR was reduced by 20.8%. Kong

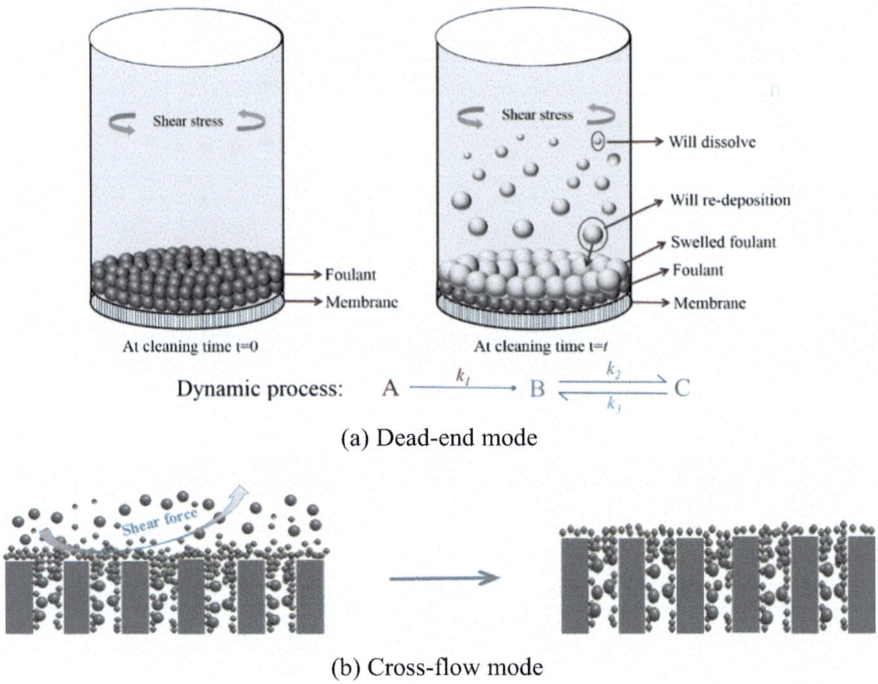

Fig. 2 Schematic diagram of the rinsing process

Fig. 3 The factors influencing on the rinsing efficiency

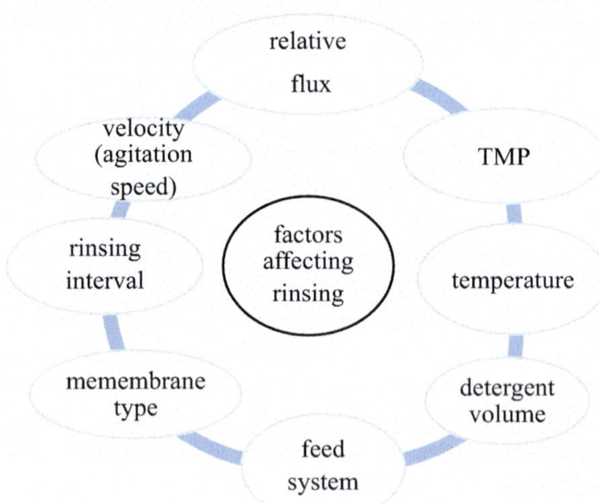

et al. [9] pointed out that the pressure in the water rinsing process decreased from 0.05 MPa to 0 MPa for the PAN membrane, and the RRNR was reduced by 18%. (3) Reducing the hardness of water can also reduce the RRNR [7, 8, 10]. (4) The RRNR in the non-circulating rinsing process is lower than that in the circulating water rinsing process (RRNR was decreased by 29%) [11]. (5) If the solubility of pollutants in water is good, the RRNR is low [12]. (6) When rinsing membranes fouled by different fouling systems, RRNR values also differ significantly: ① Wang et al. [13] pointed out that RRNR reached 29% for the membranes fouled by EPS of the activated sludge suspension in the rinsing process. ② Cabero et al. [11] pointed out that RRNR of 30% of the ceramic membrane was polluted by whey protein in the water rinsing process; Matzinos et al. [14] pointed out that RRNR was 27% for the ZrO 2 membrane polluted by whey protein the rinsing process. ③ Kong et al. [9] pointed out that RRNR is 50% for the 0.1 μm PAN membrane fouled by 1% sodium alginate solution in the rinsing process. ④ Nakanishi et al. [6] pointed out that the RRNR was 12.5% for the CA membrane polluted by skimmed milk in the rinsing process. ⑤ Rabuni et al. [15] pointed out that the RRNR was 80% for the PVDF ultrafiltration membrane fouled by bovine serum albumin (BSA) in the rinsing process.

2 Water Rinsing Model

As the most common physical cleaning method in MBR, water rinsing has the advantages of online cleaning, easy operation, and film alignment, small damage degree. However, it has the disadvantage of low cleaning efficiency (low membrane flux recovery rate), and its cleaning mechanism is unclear. Therefore, exploring the mechanism of rinsing and establishing the corresponding mathematical model will lay a theoretical foundation for the scientific development of rinsing operation.

2.1 Cabero's Model [11]

Based on a first-order model, the decrease in the membrane resistance in the water-rising process can be obtained expressed as follows:

$$-\frac{d\left(R_f - R(t)\right)}{dt} = k\left(R_f - R(t)\right) \tag{1}$$

where k is the rinsing rate constant, $R(t)$ the current resistance in the water-rising process, R_f is the membrane hydraulic resistance after fouling, and t is the time.

After integrating Eq. (1):

$$R(t) = R_r + (R_f - R_r)e^{-kt} \quad (2)$$

where R_r is the membrane hydraulic resistance after water rinsing.

The experimental results indicated that Eq. (2) only fits the final part of the rinsing curve very well. In contrast, it does not follow the experimental observations for the beginning part because the initial rinsing is faster [11].

In order to improve the accuracy of Eq. (2), a new first-order term is added to Eq. (2):

$$R(t) = R_r + (R_f - R_i)e^{-k_1 t} + (R_i - R_r)e^{-k_2 t} \quad (3)$$

where R_i is the intermediate hydraulic resistance of the membrane (at which the change in the rate constant occurs), k_1 is the rinsing rate constant for the first part of the rinsing, and k_2 is the rinsing rate constant for the second part of the rinsing.

Equation (3) showed a good agreement between model predations and experimental data ($R^2 > 0.99$) (Fig. 4). The rinsing curve is divided into two parts with different rate constants, and each part can be described in terms of first-order kinetics. Meanwhile, a faster exponential resistance decrease is observed ($k_I > k_2$) in the first stage, corresponding to loosen cake layer that is easily removed. While in the second stage, the lower rate constant corresponds to a tighter and harder cake layer. This model can estimate of the time required to reach the ultimate fouling resistance during rinsing operation in constant pressure cross-flow mode for the ceramic membrane fouled with whey protein [11].

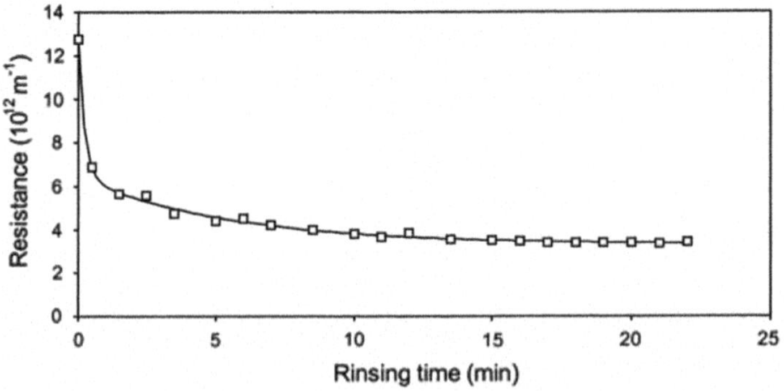

Fig. 4 Experimental data and predicted values (lines) for rinsing experiments (0.15 MPa, 2 m/s)

2.2 Kong's Empirical Model [9]

Similarly, the fouling resistance of the fouling deposition on the membrane surface in water rinsing operation can be assumed to be a first-order variation of the time derivative

$$\frac{dR_s}{dt} = k_1 R_s \tag{4}$$

After integration of Eq. (4):

$$R_s = R_s(0)e^{-k_1 t} \tag{5}$$

where Rs (0) is the fouling resistance caused by foulants deposited on the membrane surface at the end of the fouling process, m^{-1}; k_1 is the water rinsing rate constant of the removal for foulants deposited on the membrane surface, $m^{-1} \cdot s^{-1}$.

Furthermore, the foulants (sodium alginate) in the membrane pore will swell in the water rinsing process, and some small molecules or particles will enter the membrane pore under the action of shear force to increase the membrane pore resistance [16]. Therefore, the membrane pore's resistance is a time function [17]. Therefore, the change of resistance in the membrane pore follows a linear change of one variable, so the mathematical expression of this part is as follows:

$$R_{in} = k_2 t + R_{in}(0) \tag{6}$$

where k_2 is the variation rate constant of the resistance due to in-pore fouling, $m^{-1} \cdot s^{-1}$; $R_{in}(0)$ is the initial f fouling resistance due to un-swollen in-pore foulant, m^{-1} (Figs. 5 and 6).

Therefore, the total resistance in the water rinsing process can be expressed as.

$$R_t = R_s + R_{in} + R_m = R_s(0)e^{-k_1 t} + k_2 t + R_{in}(0) + R_m \tag{7}$$

Moreover, the corresponding flux can be obtained as follows:

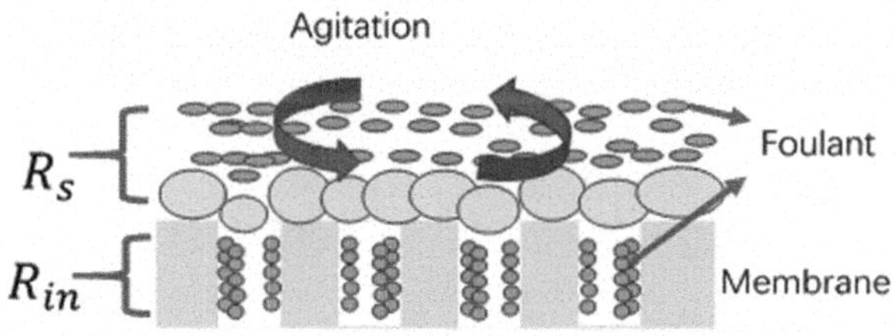

Fig. 5 Graphical representation of the water rinsing process

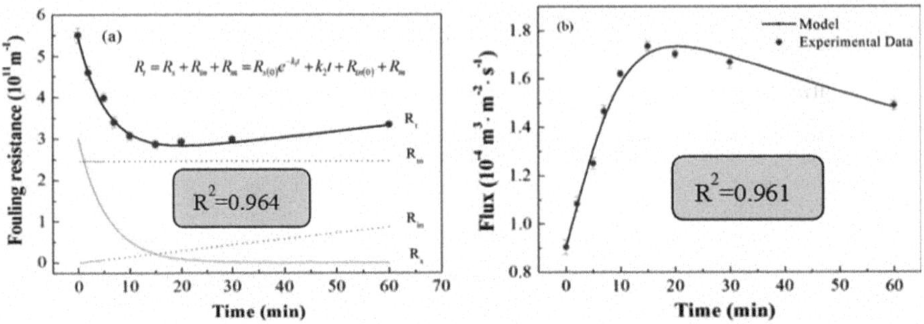

Fig. 6 Fouling resistances (**a**) and the membrane flux (**b**) in the water rinsing process at 20 °C, 0 MPa and 200 rpm (the PAN membrane, SA (sodium alginate) solution)

$$J = \frac{\Delta p}{\mu(R_s(0)e^{-k_1 t} + k_2 t + R_{in}(0) + R_m)} \tag{8}$$

The experimental results showed that the empirical model describing the fouling resistance in the water rinsing process of SA (sodium alginate)-PAN membrane system was validated, and the prediction accuracy of the proposed model was good, with a correlation coefficient greater than 0.96.

2.3 *Kong's Theoretical Model [18]*

In the water rinsing process, the foulants can be divided into the dissolution part (the part that swells first and then dissolves and the part that dissolves directly) and the swelling part (the part that does not continue to dissolve during the studied cleaning time) (Fig. 7). In addition, under shear force diffusion and Brownian diffusion, small molecular foulants on the membrane surface enter the membrane pores, causing an increase in the fouling resistance. Therefore, this part is classified as the reverse process of dissolution in this paper.

In Fig. 7, *A* is the foulant on the membrane surface and pores, and the instantaneous fouling resistance corresponding to the cleaning time *t* (s) is represented by $R_A(t)$, m^{-1}. *B* is the swelling foulant, whose instantaneous fouling resistance corresponding to the cleaning time *t* is expressed by $R_B(t)$, m^{-1}. *C* is the dissolved foulant, and its instantaneous fouling resistance at the cleaning time *t* is expressed by $R_C(t)$, m^{-1}.

Fig. 7 Schematic diagram of the water rinsing mechanism

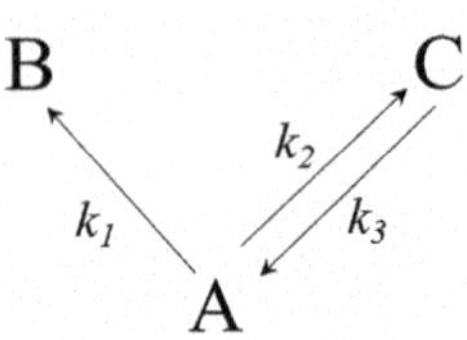

The k_1, k_2 and k_3 are the swelling, dissolution, and deposition rate constants in the water rinsing process, respectively, s^{-1}.

If the variation in the fouling resistance with rinsing time during swelling, dissolution, and deposition of pollutants conforms to first-order kinetics, the following equations can be obtained:

$$-\frac{dR_A(t)}{dt} = k_1 R_A(t) + k_2 R_A(t) - k_3 R_C(t) \tag{9}$$

$$-\frac{dR_B(t)}{dt} = -k_1 R_A(t) \tag{10}$$

$$-\frac{dR_C(t)}{dt} = -k_2 R_A(t) + k_3 R_C(t) \tag{11}$$

From Eq. (9) can be obtained:,

$$k_3 R_c(t) = \frac{dR_B(t)}{dt} + k_2 R_B(t) - R_f(0) k_1 e^{-k_1 t} \tag{12}$$

The following equation can be obtained by taking the first derivative of Eq. (12):

$$\frac{dR_C(t)}{dt} = \frac{\frac{d^2 R_A(t)}{dt^2} + (k_1 + k_2)\frac{dR_A(t)}{dt}}{k_3} \tag{13}$$

By substituting Eqs. (12) and (13) into Eq. (11) yield

$$\frac{d^2 R_A(t)}{dt^2} + (k_1 + k_2 + k_3)\frac{dR_A(t)}{dt} + k_1 k_3 R_A(t) = 0 \tag{14}$$

Equation (14) is a homogeneous linear equation ($y' + Py' + qy = 0$), after solving Eq. (14):

$$R_A(t) = c_1 e^{\lambda_1 t} + c_2 e^{\lambda_2 t} \tag{15}$$

$$\lambda_1 = \frac{-(k_1 + k_2 + k_3) + \sqrt{(k_1 + k_2 + k_3)^2 - 4k_1 k_3}}{2}$$

$$\lambda_2 = \frac{-(k_1 + k_2 + k_3) - \sqrt{(k_1 + k_2 + k_3)^2 - 4k_1 k_3}}{2}$$

At the cleaning time of 0 (s), $R_A(t) = R_f(0)$. If let $C_1 = z$ (a constant, m^{-1}), the following equations can be obtained:

$$R_A(t) = z e^{\lambda_1 t} + (R_f(0) - z) e^{\lambda_2 t}$$

$$\lambda_1 = \frac{-(k_1 + k_2 + k_3) + \sqrt{(k_1 + k_2 + k_3)^2 - 4k_1k_3}}{2}$$
$$\lambda_2 = \frac{-(k_1 + k_2 + k_3) - \sqrt{(k_1 + k_2 + k_3)^2 - 4k_1k_3}}{2} \quad (16)$$

Substituting the above Eq. (16) into Eq. (10):

$$-\frac{dR_B(t)}{dt} = -k_1\left(ze^{\lambda_1 t} + (R_f(0) - z)e^{\lambda_2 t}\right)$$
$$\lambda_1 = \frac{-(k_1 + k_2 + k_3) + \sqrt{(k_1 + k_2 + k_3)^2 - 4k_1k_3}}{2}$$
$$\lambda_2 = \frac{-(k_1 + k_2 + k_3) - \sqrt{(k_1 + k_2 + k_3)^2 - 4k_1k_3}}{2} \quad (17)$$

After integrating Eq. (17), *$R_B(t)$* can be obtained:

$$R_B(t) = \frac{zk_1}{\lambda_1}e^{\lambda_1 t} - \frac{(R_f(0) - z)k_1}{\lambda_2} - \frac{zk_1}{\lambda_1} - \frac{(R_f(0) - z)k_1}{\lambda_2} \quad (18)$$

The instantaneous fouling resistance ($R_f(t) = R_A(t) + R_B(t)$) can be expressed as follows:

$$R_f(t) = ze^{\lambda_1 t} + (R_f(0) - z)e^{\lambda_2 t} + \frac{zk_1}{\lambda_1}e^{\lambda_1 t} - \frac{(R_f(0) - z)k_1}{\lambda_2} - \frac{zk_1}{\lambda_1} - \frac{(R_f(0) - z)k_1}{\lambda_2}$$
$$\lambda_1 = \frac{-(k_1 + k_2 + k_3) + \sqrt{(k_1 + k_2 + k_3)^2 - 4k_1k_3}}{2}$$
$$\lambda_2 = \frac{-(k_1 + k_2 + k_3) - \sqrt{(k_1 + k_2 + k_3)^2 - 4k_1k_3}}{2} \quad (19)$$

where R_f *(0)* is the instantaneous fouling resistance at rinsing time of 0(s), m^{-1}.

As presented in Fig. 8, the excellent agreement between the predicted value of the proposed model and the experimental value for the instantaneous fouling resistance (a) ($R^2 = 0.980$) and the membrane flux (b) (TMP = 0 MPa, T = 20 °C) ($R^2 > 0.960$) were observed.

2.4 *Guo's Model [19]*

The water rinsing process can be considered as a composite process of the irreversible process (swelling process) and the reversible process (dissolution and deposition processes (Fig. 9):

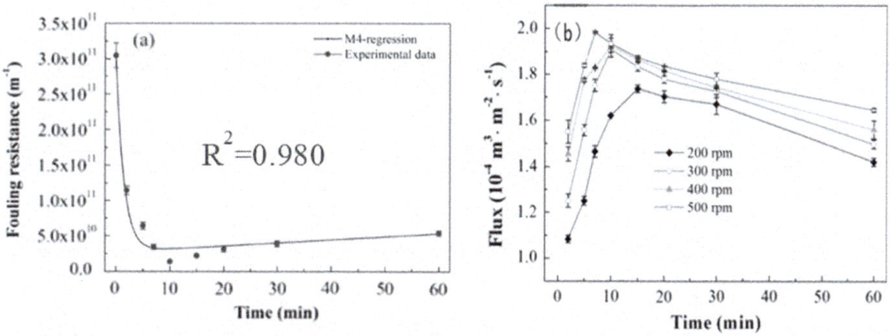

(a) the instantaneous fouling resistance (R_f(t))

(b) the membrane flux ($J(t)$)

Fig. 8 The predicted value of the proposed model and the experimental data for the 0.1 μm PAN membrane fouled with SA in the water rinsing process

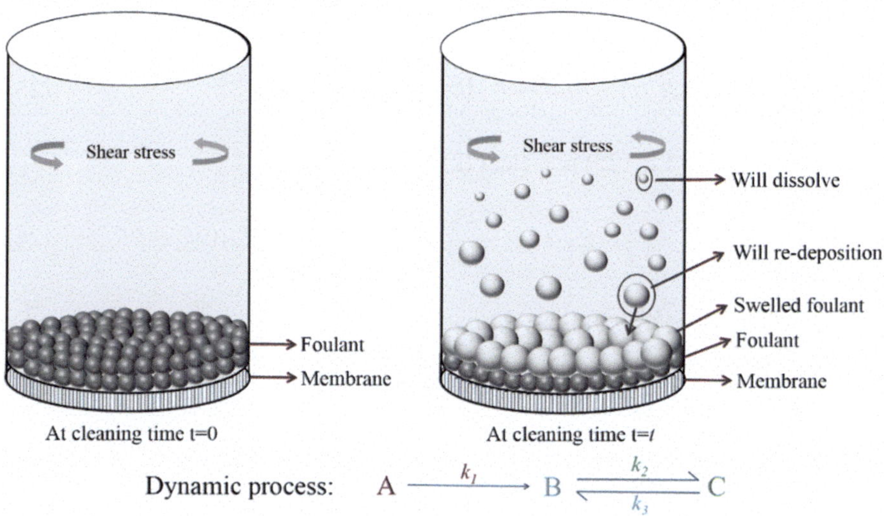

Fig. 9 The water rinsing mechanism of the fouled membrane

The A is foulants on the membrane, and its corresponding fouling resistance is R_A(t) (m^{-1}); B is the swelled foulants and its corresponding fouling resistance is R_B(t) (m^{-1}); C is the dissolved foulants and its corresponding dissolved fouling resistance is R_C(t) (m^{-1}). The k_1, k_2, and k_3 are the rate constants of swelling, dissolution, and deposition in the rinsing process (s^{-1}), respectively. The swelling, dissolution, and deposition behavior of foulants in the water rinsing process can be considered as the first-order kinetics [20–22]:

$$-\frac{dR_A(t)}{dt} = k_1 R_A(t) \quad (20)$$

$$-\frac{dR_B(t)}{dt} = -k_1 R_A(t) + k_2 R_B(t) - k_3 R_C(t) \quad (21)$$

$$-\frac{dR_C(t)}{dt} = -k_2 R_B(t) + k_3 R_C(t) \quad (22)$$

From Eq. (20)

$$R_A(t) = c_1 e^{-k_1 t} \quad (23)$$

At t = 0, $c_1 = R_A(0) = R_f(0)$, $R_A(t)$ as a function of time can be written as

$$R_A(t) = R_f(0) e^{-k_1 t} \quad (24)$$

After introducing Eq. (24) into Eq. (21):

$$k_3 R_c(t) = \frac{dR_B(t)}{dt} + k_2 R_B(t) - R_f(0) k_1 e^{-k_1 t} \quad (25)$$

After introducing Eq. (25) into Eq. (22):

$$-\frac{dR_C(t)}{dt} = -k_2 R_B(t) + \frac{dR_B(t)}{dt} + k_2 R_B(t) - R_f(0) k_1 e^{-k_1 t} \quad (26)$$

Rearranging Eq. (26):

$$\frac{dR_C(t)}{dt} = R_f(0) k_1 e^{-k_1 t} - \frac{dR_B(t)}{dt} \quad (27)$$

Derivative of Eq. (25):

$$\frac{dR_c(t)}{dt} = \frac{d^2 R_B(t)}{dt^2} \frac{1}{k_3} + \frac{R_f(0) k_1^2 e^{-k_1 t}}{k_3} + \frac{dR_B(t)}{dt} \frac{k_2}{k_3} \quad (28)$$

After introducing Eq. (28) into Eq. (27).

$$\frac{d^2 R_B(t)}{dt^2} + (k_2 + k_3) \frac{dR_B(t)}{dt} = R_f(0)(k_1 k_3 - k_1^2) e^{-k_1 t} \quad (29)$$

This is a non-homogeneous linear equation. Therefore, its solution = general solution of a homogeneous linear equation + particular solution of a non-homogeneous nonlinear equation.

The general solution of Eq. (29) is

$$R_{B1}(t) = c_2 + c_3 e^{-(k_2+k_3)t} \tag{30}$$

The following step (Eqs. (31)–(36)) is to calculate the particular solution of Eq. (29).

Suppose that the particular solution of Eq. (29) is as follows:

$$R_{B2}(t) = Ae^{-k_1 t} \tag{31}$$

$$\frac{dR_{B2}(t)}{dt} = -Ak_1 e^{-k_1 t} \tag{32}$$

$$\frac{d^2 R_{B2}(t)}{dt^2} = Ak_1^2 e^{-k_1 t} \tag{33}$$

Take Eqs. (31)–(33) into Eq. (29) to get Eq. (34).

$$Ak_1^2 e^{-k_1 t} - Ak_1(k_2 + k_3)e^{-k_1 t} = R_f(0)\left(k_1 k_3 - k_1^2\right)e^{-k_1 t} \tag{34}$$

From Eqs. (31) and (34), the particular solution of Eq. (29) as follows:

$$R_{B2}(t) = \frac{R_f(0)(k_3 - k_1)}{k_1 - k_2 - k_3} e^{-k_1 t} \tag{35}$$

From Eqs. (30) and (35), the general solution of Eq. (29) is obtained

$$R_B(t) = c_2 + c_3 e^{-(k_2+k_3)t} + \frac{R_f(0)(k_3 - k_1)}{k_1 - k_2 - k_3} e^{-k_1 t} \tag{36}$$

At t = 0, $R_B(0) = c_2 + c_3 + \frac{R_f(0)(k_3-k_1)}{k_1-k_2-k_3} = 0$, by taking c_2 equal to R_r, the constants of c_2 and c_3 can be found: $c_2 = R_r$, $c_3 = -R_r - \frac{R_f(0)(k_3-k_1)}{k_1-k_2-k_3}$. The fouling resistance after swelling as a function of cleaning time:

$$R_B(t) = \frac{R_f(0)(k_3 - k_1)}{k_1 - k_2 - k_3} e^{-k_1 t} - \left(R_r + \frac{R_f(0)(k_3 - k_1)}{k_1 - k_2 - k_3}\right) e^{-(k_2+k_3)t} + R_r \tag{37}$$

where R_r is the residual fouling resistance on the membrane (m^{-1}).

The instantaneous fouling resistance ($R_f(t) = R_A(t) + R_B(t)$) can be expressed as follows:

$$R_f(t) = R_f(0)\left(1 - \frac{k_1 - k_3}{k_1 - k_2 - k_3}\right)e^{-k_1 t} + \left(\frac{R_f(0)(k_1 - k_3)}{k_1 - k_2 - k_3} - R_r\right)e^{-(k_2+k_3)t} + R_r \tag{38}$$

According to Fig. 9, the swelling, dissolution, deposition, and net removal rate of foulants can be obtained, respectively.

$$v_s = -\frac{dR_A(t)}{dt} = k_1 R_A(t) \tag{39}$$

$$v_{di} = -\frac{dR_B(t)}{dt} = k_2 R_B(t) \tag{40}$$

$$v_{de} = -\frac{dR_C(t)}{dt} = k_3(R_f(0) - R_f(t)) \tag{41}$$

$$v_r = v_{di} - v_{de} \tag{42}$$

where v is the rate (m^{-1} s^{-1}) (subscript *s* represents swelling, *di* represents dissolution, *de* represents deposition, and *r* represents removal).

As presented in Fig. 10, the swelling rate decreased with the stirring speed (Fig. 10a–1), while the dissolution rate also dramatically varied with the stirring speed (Fig. 10a–2). Besides, the deposition rate of foulants also increased with stirring speed (Fig. 10a–3) due to the dynamic equilibrium established by the dissolution process and the deposition process. In terms of order of magnitude, the swelling rate is about 10 times the dissolution rate and 100 times the deposition rate. Moreover, the proposed model can be widely used to predict the instantaneous fouling resistance of microfiltration/ultrafiltration membrane fouled with AS, SA, BSA, and HA in the hydraulic cleaning process, and the good agreement between the model predictions and the experimental data for the instantaneous fouling resistance in long-term experiment (four filtrations and cleaning cycles) was also observed (Fig. 10b) (R^2 was 0.9765 (1st water rinsing cycle), 0.9974 (2nd water rinsing cycle), 0.9948 (3rd water rinsing cycle) and 0.9993 (4th water rinsing cycle), respectively). The initial fouling resistance ($R_f(t = 0)$) increased with the cycle filtration and this is because some irreversible foulants were difficult to remove in the water rinsing process [23].

2.5 *Wang's Model [24]*

Cake and gel are two common forms of foulants on the membrane surface [25, 26]. The former is easier to remove than the latter under in the rinsing process under shear stress due to their different natures and interaction forces. In contrast, the former (gel) must first swell to reduce the attractive interaction energy to satisfy the following gel dissolution. Meanwhile, some removed foulants in the rising process will re-deposit on the membrane surface and into pores [27]. The dynamics change of gel and cake in the hydraulic cleaning process can be described by the following schematic diagram (Fig. 11):

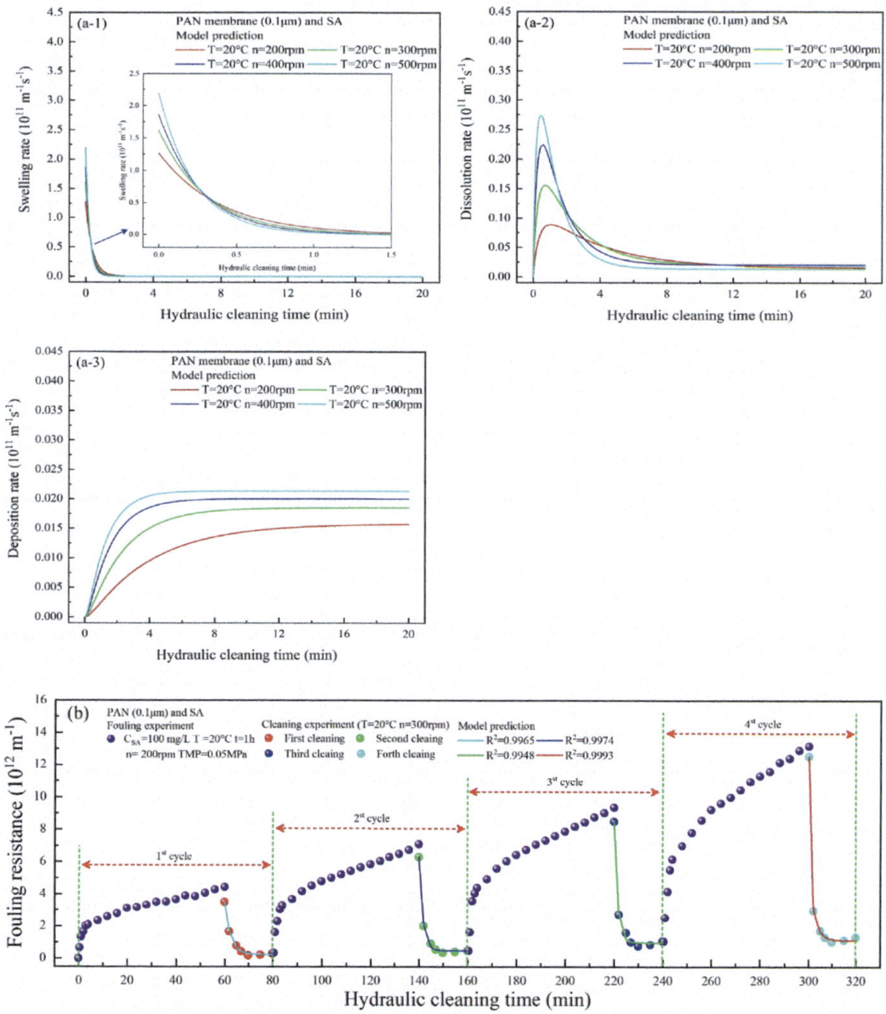

Fig. 10 Model validation. The effects of temperature on the swelling, dissolution, and deposition rates of foulants (**a**) and in the cyclic cleaning (**b**). 0.1 μm PAN fouled by the activated sludge (AS)

In the water rinsing process, the instantaneous fouling resistance for the virgin foulants (the sum of gel and cake) (*A*), the swelled gel (*B*), the directly dissolved cake (*C*), and the dissolved gel (*D*), respectively, are R_A (t), R_B (t), R_C (t), and R_D (t) (m^{-1}). The swelled gel means that water molecules have entered the gel in the water rinsing process and caused its volume to expand. The re-deposition refers to the removed gel and cake, which will back-transfer to the membrane surface and deposit on the membrane surface or in pores under the action of shear stress in the water rinsing process.

Fig. 11 The water rinsing mechanism of the fouled membrane

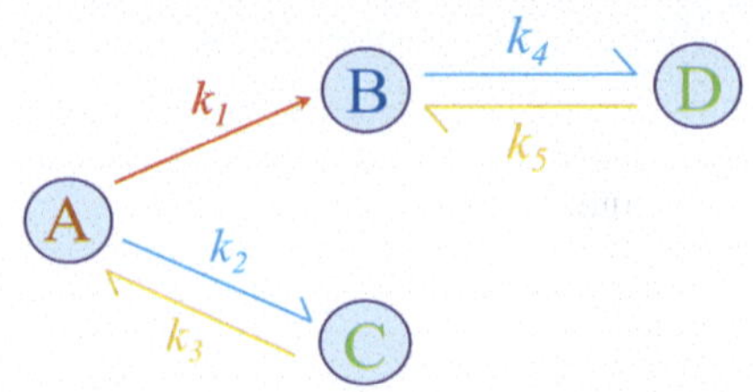

A virgin foulants (gel and cake) B swollen foulants

C dissolved cake D dissolved gel

k_1 (s^{-1}) is the rate constant of swelling (gel)

k_2 (s^{-1}) is the rate constant of dissolution (cake)

k_3 (s^{-1}) is the rate constant of re-deposition (dissolved cake)

k_4 (s^{-1}) is the rate constant of dissolution (swollen gel)

k_5 (s^{-1}) is the rate constant of re-deposition (dissolved gel)

According to the schematic diagram (Fig. 11), the corresponding kinetics equations can be expressed as follows:

$$-\frac{dR_A(t)}{dt} = k_1 R_A(t) + k_2 R_A(t) - k_3 R_C(t) \tag{43}$$

$$-\frac{dR_B(t)}{dt} = -k_1 R_A(t) + k_4 R_B(t) - k_5 R_D(t) \tag{44}$$

$$-\frac{dR_C(t)}{dt} = -k_2 R_A(t) + k_3 R_C(t) \tag{45}$$

$$-\frac{dR_D(t)}{dt} = -k_4 R_B(t) + k_5 R_D(t) \tag{46}$$

where k_1 (s^{-1}) is the rate constant of swelling (gel), k_2 (s^{-1}) is the rate constant of dissolution (cake), k_3 (s^{-1}) is the rate constant of re-deposition (dissolved cake), k_4 (s^{-1}) is the rate constant of dissolution (swelled gel) and k_5 (s^{-1}) is the rate constant of re-deposition (dissolved gel), respectively.

Combining Eq. (45) into Eq. (43) to get

$$k_3 R_C(t) = k_1 R_A(t) v + k_2 R_A(t) + \frac{dR_A(t)}{dt} \tag{47}$$

After the derivative of Eq. (47):

$$\frac{dR_C(t)}{dt} = \frac{(k_1 + k_2)}{k_3}\frac{R_A(t)}{dt} + \frac{1}{k_3}\frac{d^2R_A(t)}{dt^2} \tag{48}$$

Substitute Eqs. (47) and (48) into Eq. (45)

$$\frac{d^2R_A(t)}{dt^2} + (k_1 + k_2 + k_3)\frac{R_A(t)}{dt} + k_1k_3R_A(t) = 0 \tag{49}$$

After solving the homogeneous differential Eqs. (49), and (50) can be obtained:

$$R_A(t) = C_1e^{\lambda_1 t} + C_2e^{\lambda_2 t} \tag{50}$$

where

$$\lambda_1 = \frac{-(k_1 + k_2 + k_3) + \sqrt{(k_1 + k_2 + k_3)^2 - 4k_1k_3}}{2}$$
$$\lambda_2 = \frac{-(k_1 + k_2 + k_3) + \sqrt{(k_1 + k_2 + k_3)^2 - 4k_1k_3}}{2}$$

$R_A(t) = R_f\ (0)$ at t = 0, order $C1 = R_r$, substitution formula (50):

$$R_A(t) = R_re^{\lambda_1 t} + \left(R_f(0) - R_r\right)e^{\lambda_2 t} \tag{51}$$

where, $R_f(0)$ is the fouling resistance at the cleaning time of $0(\text{m}^{-1})$; R_r is the residual foluing resistance on the membrane (m^{-1}).

Transformation of Eq. (44) leads to Eq. (52)

$$k_5R_D(t) = -k_1R_A(t) + k_4R_B(t) + \frac{dR_B(t)}{dt} \tag{52}$$

By taking the first derivative of Eq. (52):

$$\frac{k_5R_D(t)}{dt} = -\frac{k_1R_A(t)}{dt} + \frac{k_4R_B(t)}{dt} + \frac{d^2R_B(t)}{dt^2} \tag{53}$$

Substitute Eqs. (51), (52) and (53) into Eq. (46) to get

$$\frac{d^2R_B(t)}{dt^2} + (k_4 + k_5)\frac{R_B(t)}{dt} = k_1(k_5 + \lambda_1)R_re^{\lambda_1 t} + k_1(k_5 + \lambda_2)\left(R_f(0) - R_r\right)e^{\lambda_2 t} \tag{54}$$

After solving the non-homogeneous differential equation of (54):

$$R_B(t) = R_r \frac{k_1 + k_1 k_5/\alpha}{\alpha + k_4 + k_5}\left(e^{\alpha t} - 1\right) + \left(R_f(0) - R_r\right)\frac{k_1 + k_1 k_5/\beta}{\beta + k_4 + k_5}\left(e^{\beta t} - e^{-(k_4+k_5)t}\right) \tag{55}$$

By introducing Eq. (51) into Eq. (45), we can get

$$\frac{dR_C(t)}{dt} + k_3 R_C(t) = k_2\left(R_r e^{\alpha t} + \left(R_f(0) - R_r\right)e^{\beta t}\right) \tag{56}$$

The fouling resistance ($R_c(t)$) of the dissolved filter cake can be obtained by the integral solution of Eq. (56).

$$R_C(t) = \frac{k_2 R_r}{\alpha + k_3}\left(e^{\alpha t} - e^{-k_3 t}\right) + \frac{k_2\left(R_f(0) - R_r\right)}{\beta + k_3}\left(e^{\beta t} - e^{-k_3 t}\right) \tag{57}$$

At any cleaning time, the sum of the fouling resistance ($R_A(t) + R_B(t)$) and dissolved fouling resistance ($R_C(t) + R_D(t)$) is $R_f(0)$. Therefore, the fouling resistance ($R_D(t)$) of dissolved gel can be expressed as

$$R_D(t) = R_f(0) - R_A(t) - R_B(t) - R_C(t) \tag{58}$$

The instantaneous pollution resistance ($R_f(t)$) at cleaning time t:

$$\begin{aligned} R_f(t) = {} & R_r\left(e^{\alpha t} + \frac{k_1 + k_1 k_5/\alpha}{\alpha + k_4 + k_5}\left(e^{\alpha t} - 1\right)\right) \\ & + \left(R_f(0) - R_r\right)\left(e^{\beta t} + \frac{k_1 + k_1 k_5/\beta}{\beta + k_4 + k_5}\left(e^{\beta t} - e^{-(k_4+k_5)t}\right)\right) \end{aligned} \tag{59}$$

And the corresponding membrane flux in the water rinsing process can be expressed as

$$J(t) = \frac{\Delta p}{\mu\left(R_m + R_r\left(e^{\alpha t} + \frac{k_1 + k_1 k_5/\alpha}{\alpha + k_4 + k_5}\left(e^{\alpha t} - 1\right)\right) + \left(R_f(0) - R_r\right)\left(e^{\beta t} + \frac{k_1 + k_1 k_5/\beta}{\beta + k_4 + k_5}\left(e^{\beta t} - e^{-(k_4+k_5)t}\right)\right)\right)} \tag{60}$$

It can be seen in Fig. 12, the proposed model can be used to predict the variations of the instantaneous fouling resistance of gel and cake ($R_A(t)$), swollen gel ($R_B(t)$), and dissolved gel and cake ($R_{C+D}(t)$) in the hydraulic cleaning process (a). Meanwhile, good agreements were observed for model predictions and experimental data in the cycle cleaning process ($R^2 > 0.9875$) (b) and the proposed model also showed a good prediction for other UF/MF membranes (c).

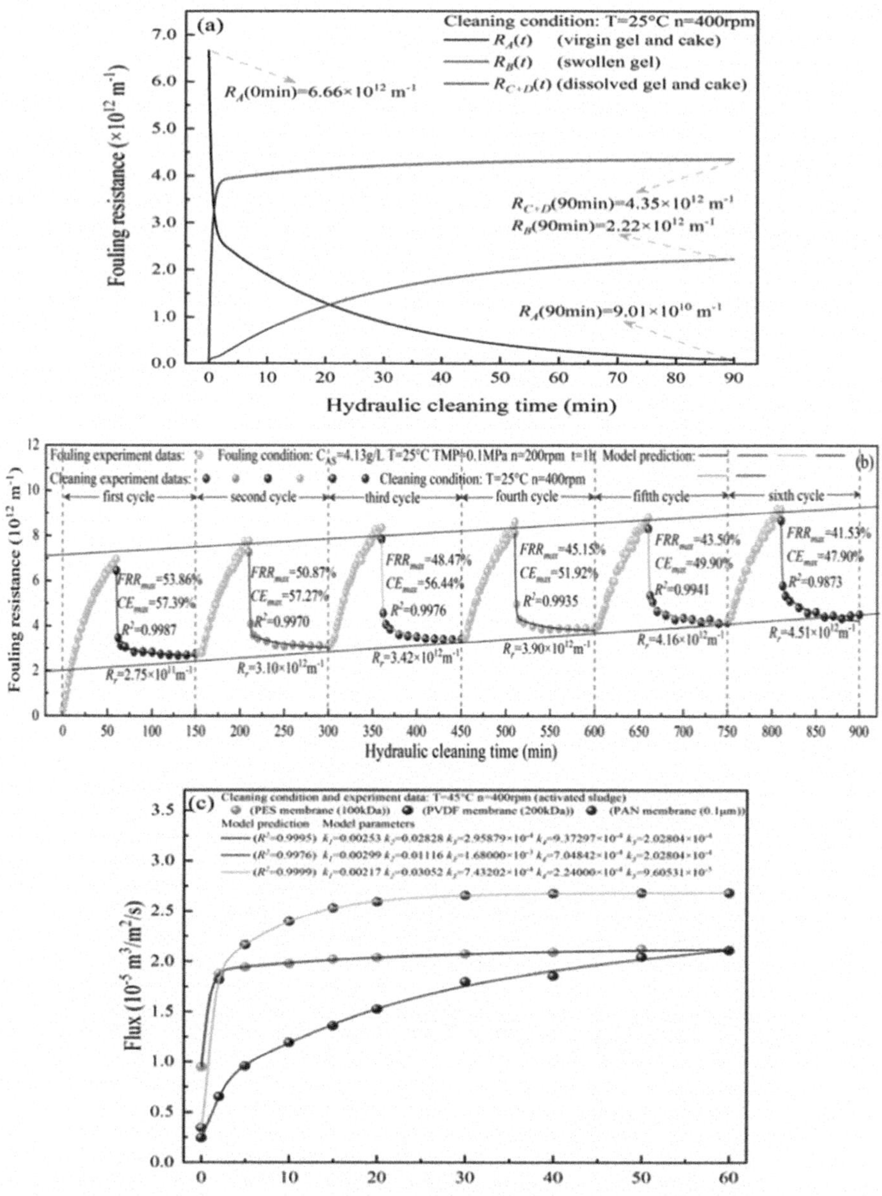

Fig. 12 The instantaneous fouling resistance of gel and cake ($R_A(t)$), swollen gel ($R_B(t)$) and dissolved gel and cake ($R_{C+D}(t)$) in the hydraulic cleaning process(a). The comparison of model predictions and experimental data in the cycle cleaning process for R_f(t) at temperature of 25 °C, stirring speed of 400 rpm(b), and the application of the proposed model for different fouled membranes with different pore sizes (c)

2.6 Qiao's Model [28]

As Cabero et al. [11] reported, the foulants on the membrane surface can be divided into the loose foulants and the compact foulants in the ringing process, which indicates that the cake on the membrane can also be divided into loose and compact cakes in the cleaning process. Meanwhile, the particle size of some foulants will be reduced under the action of shear stress, and then these foulants with the smaller particle size will be re-deposited in the pores of the membrane and cake.

The total fouling resistance at the rinsing time of t can be expressed as

$$R_f(t) = R_{loose}(t) + R_{compact}(t) \tag{61}$$

Suppose that the removal of loose cake and compact cake in the rinsing process can be, respectively, expressed by the first-order kinetic equation:

$$\frac{dR_{loose}(t)}{dt} = -k_1 R_{loose}(t) \tag{62}$$

$$\frac{dR_{compact}(t)}{dt} = -k_2 R_{compact}(t) \tag{63}$$

After integrating Eqs. (62) and (63):

$$R_{loose}(t) = R_{loose}(0)e^{-k_1 t} \tag{64}$$

$$R_{compact}(t) = R_{compact}(0)e^{-k_2 t} \tag{65}$$

After inserting Eqs. (64) and (65) into Eq. (61):

$$R_f(t) = R_{loose}(0)e^{-k_1 t} + R_{compact}(0)e^{-k_2 t} \tag{66}$$

Considering the fact that some rinsed foulants can be re-deposited in the pores of the membrane and cake, Eq. (66) needs to be modified as

$$\begin{aligned} R_f(t) = {} & R_{loose}(0)e^{-k_1 t} + R_{compact}(0)e^{-k_2 t} \\ & + \left(R_f(0) - R_{loose}(0)e^{-k_1 t} - R_{compact}(0)e^{-k_2 t}\right)k_3 t \end{aligned} \tag{67}$$

where k_3 is the re-deposition rate constant of foulants (s^{-1}).

In addition, the removed fouling resistance of the loose cake ($R_{r\text{-}loose}(t)$) and the compact cake ($R_{r\text{-}compact}(t)$), and the re-deposition resistance of foulants ($R_{re\text{-}deposition}(t)$) ($m^{-1}$) can be, respectively, obtained as

$$R_{r-loose}(t) = R_{loose}(0) - R_{loose}(0)e^{-k_1 t} \tag{68}$$

$$R_{r\text{-}compact}(t) = R_{compact}(0) - R_{compact}(0)e^{-k_2 t} \tag{69}$$

$$R_{re-deposition}(t) = (R_f(0) - R_{loose}(0)e^{-k_1 t} - R_{compact}(0)e^{-k_2 t})k_3 t \tag{70}$$

Furthermore, the corresponding removal rate of the loose cake ($r_{r\text{-}loose}(t)$) and the compact cake ($r_{r\text{-}compact}(t)$), and the re-deposition rate of foulant ($r_{re\text{-}deposition}(t)$) ($m^{-1}\ s^{-1}$) at the rinsing time of t can be, respectively, expressed as

$$r_{r\text{-}loose}(t) = \frac{dR_{r\text{-}loose}(t)}{dt} = k_1 R_{loose}(0)e^{-k_1 t} \tag{71}$$

$$r_{r-compact}(t) = \frac{dR_{r-compact}(t)}{dt} = k_2 R_{compact}(0)e^{-k_2 t} \tag{72}$$

$$r_{re-deposition}(t) = \frac{dR_d(t)}{dt} = \left(R_f(0) + R_{loose}(0)(k_1 t - 1) + R_{compact}(0)(k_2 t - 1)\right)k_3 \tag{73}$$

Therefore, the water rinsing process can be divided into three stages (Fig. 13): (i) the first stage: the loose cake was quickly removed, resulting in the rapid decrease of $R_f(t)$, and this process had been completed at the first transition point (t_1), (ii) the second stage: the compact cake was difficult to be removed, leading to the slow decrease of $R_f(t)$ and the removal rate of the compact cake ($r_{r\text{-}compact}$) was equal to the re-deposition rate of foulants ($r_{re\text{-}deposition}$) at the second transition point (t_2) ($t_2 > t_1$), (iii) the third stage: $r_{r\text{-}compact}$ was smaller than $r_{re\text{-}deposition}$ after t_2, resulting in the slight increase of $R_f(t)$.

The model predictions had good agreements with experimental data for the different membranes fouled with AS suspension under the same water rinsing conditions (25 °C and 2.5 m/s) ($R^2 > 0.9990$) (Fig. 14a) and in the cycle rinsing process of the different water rinsing conditions (Fig. 14b).

2.7 *Yao's Model [29]*

In the water rinsing process, the detachment of foulants from the membrane under hydrodynamic shear force is based on a mechanism known as "peeling" or "rolling" phenomenon [30] and the shear force would strongly increase the "leverage" between particles and the planar surface to facilitate detachment [31]. However, it is difficult to measure the "leverage" in practice.

The detachment of foulants from the membrane under hydrodynamic shear force in the water rinsing process is based on a mechanism known as the "peeling" or

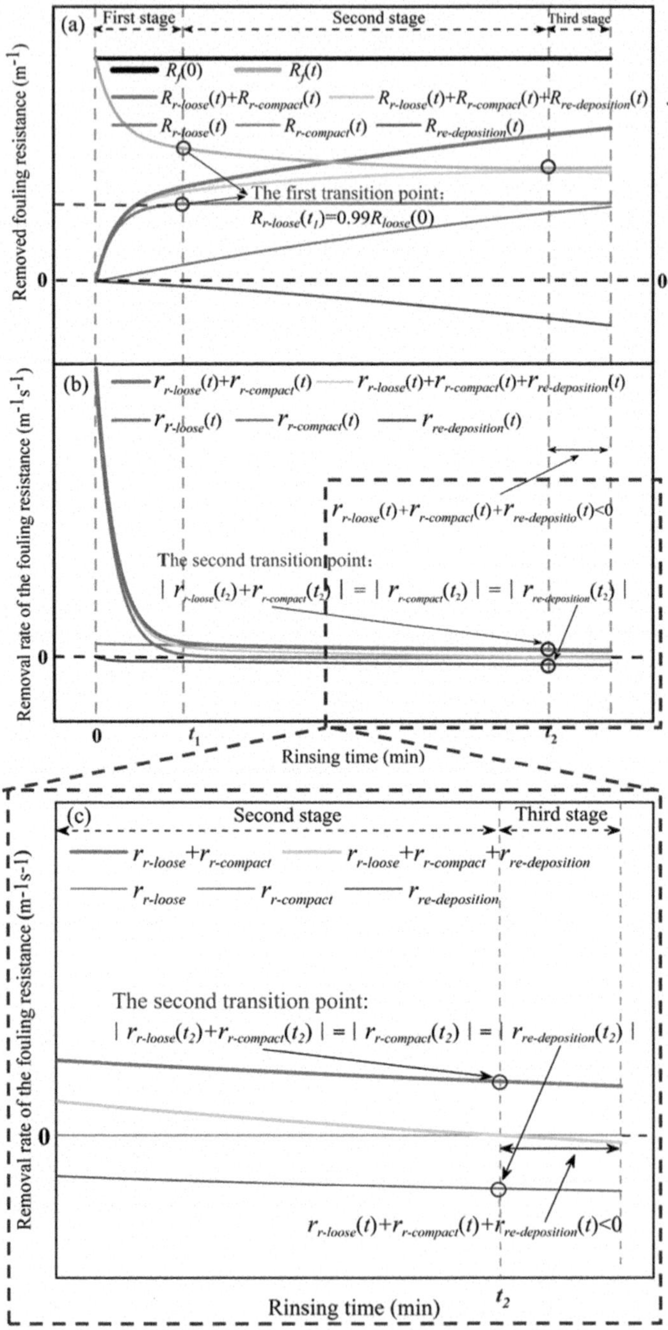

Fig. 13 Determination of the transition point of three different stages

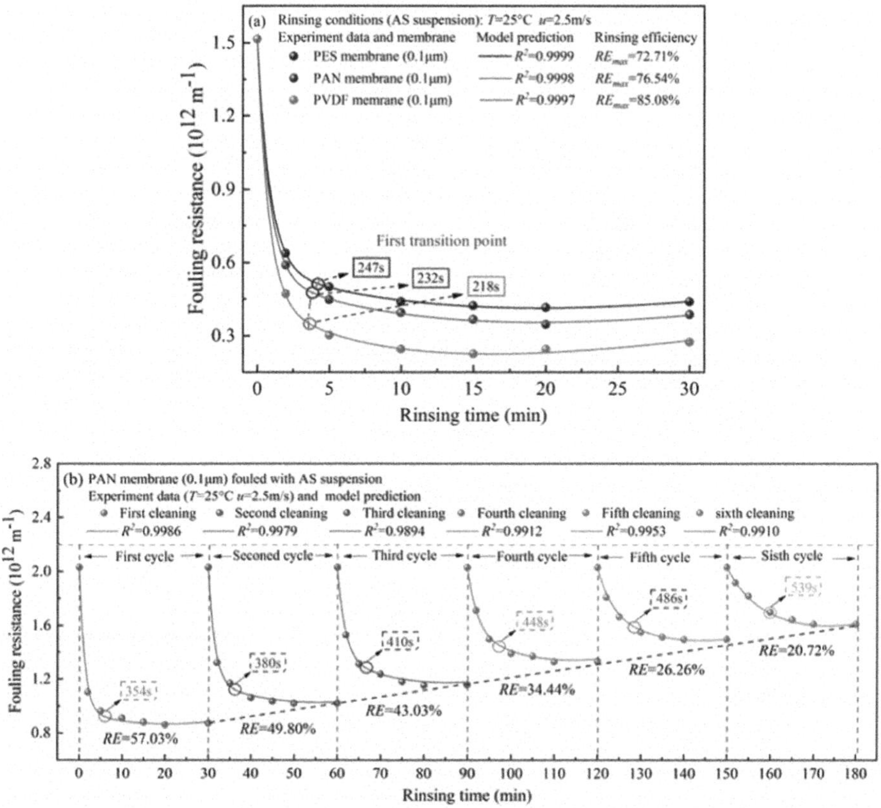

Fig. 14 The comparison of model predictions with experimental data under the same water rinsing conditions (25 °C and 2.5 m/s). **a** Different MF membranes (0.1 μm PES, PAN, and PVDF MF membranes) and **b** in the cycle rinsing process

"rolling" phenomenon [30]. In addition, the shear force strongly increases the "leverage" between particles and the planar surface to facilitate detachment [31]. However, it is difficult to measure the "leverage" in practice.

If linked the "leverage" (L) with the separation distance (h) between particles and the planar surface. L increases with the increment of h as particles roll over the planar surface (Fig. 15). Furthermore, they have a linear relationship ($r + h = L\cos\theta$). Therefore, the essence of the water rinsing process can be considered as prolonging the separation distance between foulant particles and planar surfaces.

The water rinsing process destroys the interaction energy between foulant particles and a planar surface. In order to clarify the relationship between interaction energy and energy consumption, the black-box theory is used (Fig. 16):

1. Before the fouling process and after the water rinsing process, the foulant particles keep in a free state;

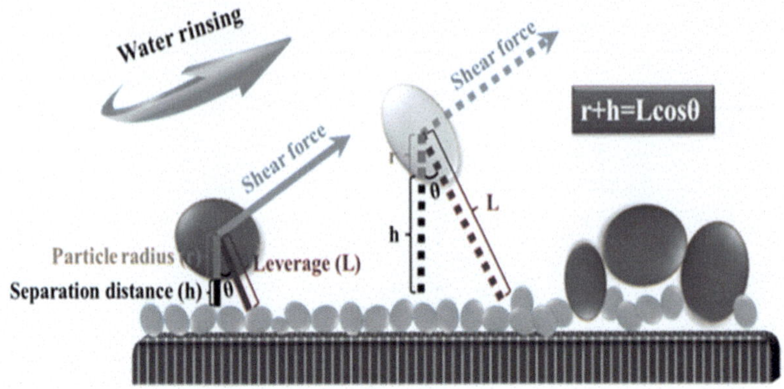

Fig. 15 Mechanism of the water rinsing process

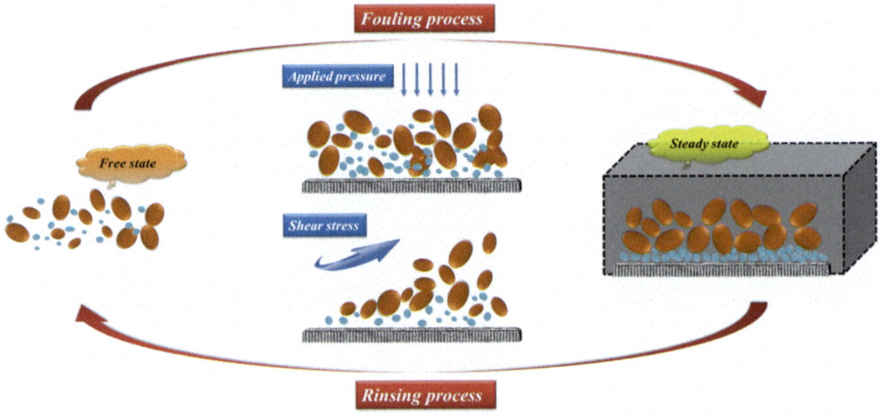

Fig. 16 Schematic diagram of the black-box theory for the energy consumption in the fouling and the water rinsing processes

2. The cake layer is affected by the external force (pressure or shear stress) in the fouling and water rinsing processes;
3. Without an external force, whatever the cake layer is swelling or stretching, it remains stable in the suspension (black box).

From the energy balance view, the energy consumption for sludge particles breaks away from the planar surface and return to the free state (rinsing process) should be equal to the energy requirement to form the cake layer (fouling process), i.e., the energy barrier, which is mainly used as the maximum interaction energy to overcome for particles adhere to the planar surface, is now acted as the minimum energy required for particles release from the surface.

Once the energy consumption as particles rolling over the unique planar surface is determined, variations of energy consumption in the water rinsing process can also

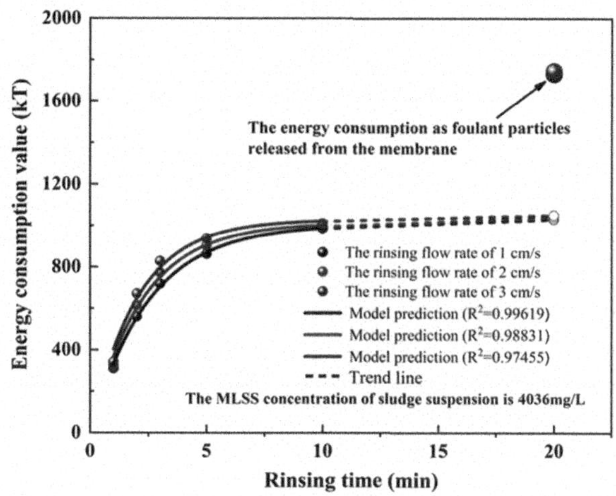

Fig. 17 Profiles of the variation of the energy consumption at different rinsing time and model predictions (the feed suspension was taken from SBR1, the rinsing flow rate is 1, 2, and 3 cm/s, respectively, and both the membrane and the surface of the flaunt layer were rough)

be confirmed. For example, suppose that the flouant layers are washed layer by layer from top to bottom in the water rinsing process and the interaction energy between foulant layers in the cake and the shear stress of the rinsing flow are two primary factors for the variation of the energy consumption in the water rinsing process [11, 32], a first-order kinetic equation of energy consumption can be established to describe the water rinsing process:

$$\eta = \delta_1\left[1 - \exp(-\delta_2 t)\right] \tag{74}$$

where the energy consumption (η, kT) at different rinsing time is determined by the characteristic parameter of interaction energy for specific sludge suspension (δ_1) (kT) and the parameter of the detachment rate (δ_2) (s^{-1}).

The model predictions fit well with the experimental data ($R^2 > 0.97$) (Fig. 17). The energy consumption value in the water-washing process increases with the increase in cleaning time and tends to be stable at 10 min. The higher the water-washing flow rate, the faster the energy consumption value reaches stability.

2.8 Qao's Model [33]

In the water singing process, when the cross-flow lift (F_{CL}) produced by the cross-flow velocity is more significant than the interaction force (F_{IF}) between foulants and foulants/the membrane surface (F_{IF} is a function of separation distance [34]), the foulants will begin to be dragged away from the membrane surface and then the foulants can be completely removed until the separation distance reaches a specific separation distance (h_{max}) (i.e., $F_{IF} = 0$ nN). Meanwhile, the cake porosity decreases gradually from top to bottom [29, 35] while the corresponding F_{IF} on the membrane surface will increase and the further removal of foulants is gradually becoming more difficult.

The cake porosity (ε_c) on the membrane surface:

$$\varepsilon_c = 1 - \frac{V_p}{V} = 1 - \frac{\frac{4}{3}\pi \cdot a_p^3 \cdot N}{SL} \tag{75}$$

where V_p is the total volume of foulant particles (m^3), V is the cake volume (m^3), a_p is the particle radius (m), N is the particle number of foulants on the membrane surface, S is the membrane surface area (m^2), and L is the cake thickness (m).

According to the water rinsing mechanism in Fig. 18, the particle number of foulants/cake thickness as a function of cross-flow velocity and water rinsing time can be expressed as [36]

$$\frac{dN}{d(u - u_0)} = -k_1 N t \tag{76}$$

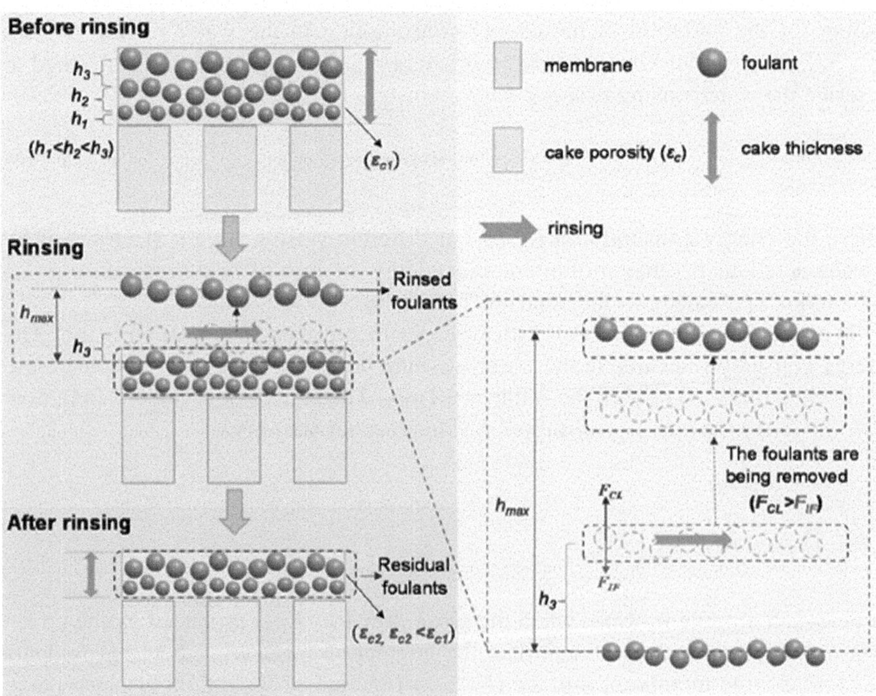

Fig. 18 The removal mechanism of foulants in the rinsing process

$$\frac{dL}{d(u-u_0)} = -k_2 L t \tag{77}$$

where t is the water rinsing time (s), k_1 is the removal rate constant of foulant particles (s^{-1}), u is the cross-flow velocity (m/s), u_0 is the minimum cross-flow velocity required (m/s), and k_2 is the variation rate constant of cake thickness (s^{-1}).

After integration:

$$N = N_0 \mathrm{e}^{-k_1(u-u_0)t} \tag{78}$$

$$L = L_0 \mathrm{e}^{-k_2(u-u_0)t} \tag{79}$$

where N_0 is the initial particle number of foulants and L_0 is the initial cake thickness (m).

After inserting Eqs. (78) and (79) into Eq. (75):

$$\varepsilon_c = 1 - \frac{\frac{4}{3}\pi \cdot a_p^3 \cdot N_0 \cdot e^{-k_1(u-u_0)t}}{S L_0 \mathrm{e}^{-k_2(u-u_0)t}} \tag{80}$$

According to the Kozemy-Carman equation [37], the specific cake resistance (r, m^{-2}) can be expressed as

$$r = \frac{180(1-\varepsilon_c)^2}{(2a_q)^2 \varepsilon_c^3} \tag{81}$$

The instantaneous fouling resistance ($R(t)$, m^{-1}) can be obtained:

$$R(t) = R_0 - \int_L^{L_0} r\,dL \tag{82}$$

where R_0 is the initial fouling resistance (m^{-1}).

After inserting Eqs. (80) and (81) into Eq. (82):

$$R(t) = R_0 - \frac{30\pi \cdot a_p \cdot N_0 \cdot e^{-k_1(u-u_0)t}}{S} \cdot \left[\frac{1}{\left(1 - \frac{4\pi \cdot a_p^3 \cdot N_0 \cdot e^{-k_1(u-u_0)t}}{3 S L_0 e^{-k_2(u-u_0)t}}\right)^2} - \frac{1}{\left(1 - \frac{4\pi \cdot a_p^3 \cdot N_0 \cdot e^{-k_1(u-u_0)t}}{3 S L_0}\right)^2} \right] \tag{83}$$

The membrane rinsing efficiency (η) was given by

$$\eta = \left(1 - \frac{R(t)}{R_0}\right) \times 100\% = \frac{30\pi \cdot a_p \cdot N_0 \cdot e^{-k_1(u-u_0)t}}{SR_f(0)} \cdot \left[\frac{1}{\left(1 - \frac{4\pi \cdot a_p^3 \cdot N_0 \cdot e^{-k_1(u-u_0)t}}{3SL_0e^{-k_2(u-u_0)t}}\right)^2} - \frac{1}{\left(1 - \frac{4\pi \cdot a_p^3 \cdot N_0 \cdot e^{-k_1(u-u_0)t}}{3SL_0}\right)^2}\right] \times 100\% \tag{84}$$

The instantaneous fouling resistances of the rinsed loose cake ($R(t)_{loose}$, m^{-1}) and compact cake ($R(t)_{compact}$, m^{-1}) can be calculated by the following equations, respectively.

$$R(t)_{compact} = R_0 - R - R_{loose} \tag{85}$$

$$R(t)_{compact} = R_0 - R - R_{loose} \tag{86}$$

In addition, the energy consumption (W, J) for the cake removal in the water rinsing process can be calculated as

$$W = \sum_{0}^{n}\left[\frac{2F_{IF}h_{\max} - \int_0^{h_n} F_{IF}dh - \int_0^{h_{n-1}} F_{IF}dh}{2} \times (N_n - N_{n-1})\right] \tag{87}$$

where n is the number of times that η increases by a certain value; h_{max} is the separation distance when the interaction force between foulants and foulants/the membrane surface is 0 nN (nm). h_n is the separation distance between the rinsed foulants and the foulants/membrane surface when η increases by a certain value for n times (nm) and N_n is the particle number of rinsed foulants when η increases by a certain value for n times.

The model predictions were in good agreement with experimental data ($R^2 > 0.9640$) (Fig. 19a). However, the corresponding maximum of the rinsing efficiency (η_{max}) gradually decreased from 76.54% (1st cycle) to 50.39% (5th cycle) and u_η (the velocity at transition point corresponding to optimal W) decreased from 2.00 m/s (1st cycle) to 1.78 m/s (5th cycle) with the increase of cycle. It was mainly because the smaller foulant particles were forced into the gap of the remaining cake and accumulated to form the cake in the subsequent filtration process [38], which led to the decrease of L_0 (L at 0.7 m/s) (Fig. 19b) and ε_c (Fig. 19c) with the circulating filtrations. However, the corresponding r increased gradually after each filtration cycle (Fig. 19d). As a result, F_{IF} gradually increased [9] and the removal of the cake became more difficult with the execution of multiple filtration and rinsing processes. Meanwhile, W tended to increase with the increase of cycle (Fig. 19e), which further indicated that the greater interaction energy needed to be overcome to achieve the same η and the removal of foulants became more difficult with the increase of cycle [39]. For example, W increased from 2.72×10^{-11} J (1st cycle) to 6.08×10^{-11} J

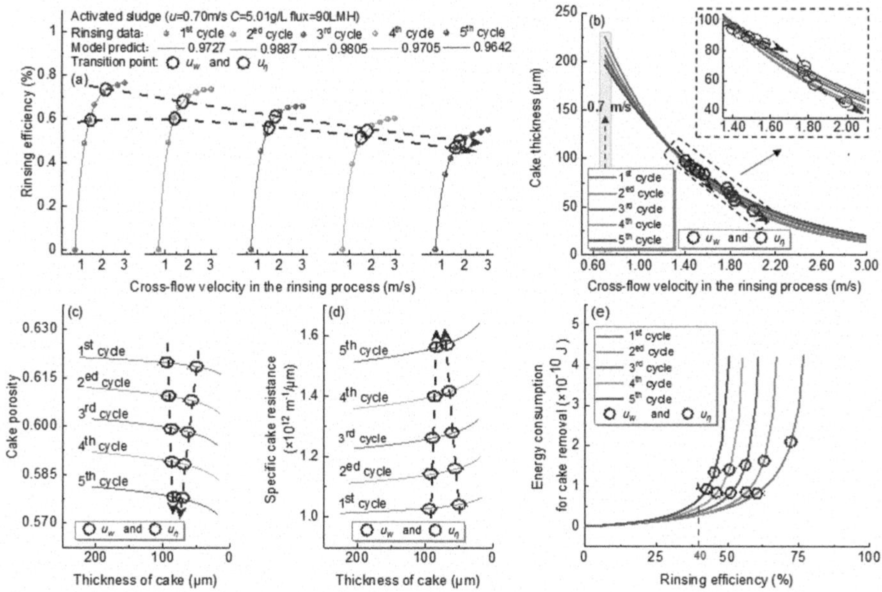

Fig. 19 Model prediction of a 0.1 μm PAN membrane fouled with activated sludge suspension in the multiple filtration and rinsing processes. **a** rinsing efficiency, **b** cake thickness, **c** cake porosity, **d** specific cake resistance, and **e** energy consumption for the cake removal

(5th cycle) at η of 40%. In addition, u_w ((corresponding to optimal W)) increased gradually from 1.40 m/s (1st cycle) to 1.56 m/s (5th cycle) and the suitable cross-flow velocity range from u_w to u_η (the transition point corresponding to optimal η) at a fixed temperature was gradually narrowing with the increase of cycle. Therefore, for the residual foulants that cannot be removed in the long-term filtration and rinsing cycles, other methods (backwash or chemical backwash) were more suitable for the further removal of foulants to reduce W and improve η.

2.9 Wang's Model [40]

Cabero's Model is established based on the first-order dynamics of the fouling resistance change with time in the water rinsing process [11]. However, in the actual cleaning process, the resistance change with time may be of order m, and m maybe 0.5, 1 or more than 1. In this case, the following cascade model with the m order in the second term and with n order in the third term can be established [40]:

$$R_f(t) = R_r + \frac{R_{f0} - R_r}{\sqrt[m-1]{1 + (m-1)(R_{f0} - R_i)^{m-1} k_1 t}^{\,m}} + \frac{R_i - R_r}{\sqrt[n-1]{1 + (n-1)(R_i - R_r)^{n-1} k_2 t}} \tag{88}$$

Table 1 Specific mathematical expressions for cascade kinetic models

Kinetic model		Reaction order	Mathematical expression
Cascade models	M_{1+1}	$1+1$	$R = R_r + (R_{f0} - R_i)e^{-k_1 t} + (R_i - R_r)e^{-k_2 t}$
	M_{1+2}	$1+2$	$R = R_r + (R_{f0} - R_i)e^{-k_1 t} + \frac{R_i - R_r}{1+(R_i - R_r)k_2 t}$
	M_{2+1}	$2+1$	$R = R_r + \frac{R_{f0} - R_i}{1+(R_{f0} - R_i)k_1 t} + (R_i - R_r)e^{-k_2 t}$
	M_{2+2}	$2+2$	$R = R_r + \frac{R_{f0} - R_i}{1+(R_{f0} - R_i)k_1 t} + \frac{R_i - R_r}{1+(R_i - R_r)k_2 t}$

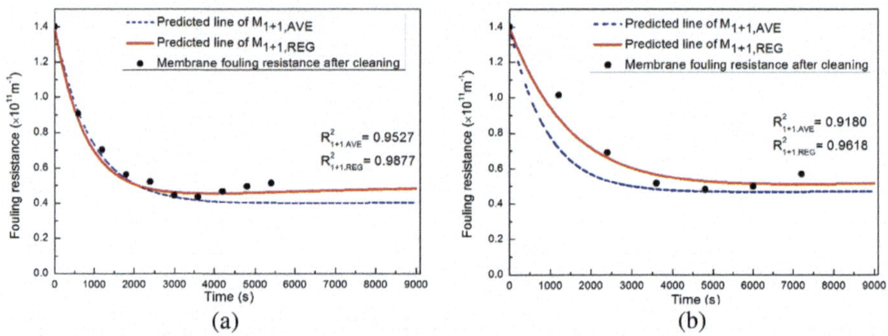

Fig. 20 Comparisons of the predictions of $M_{1+1,REG}$ or $M_{1+1,AVE}$ with the experimental data at different cleaning conditions: **a** C = 200 mM, ω = 600 rpm, T = 25°C; **b** C = 100 mM, ω = 600 rpm, T = 15 °C

When *m* and *n* take different numbers, the corresponding specific mathematical expressions for cascade kinetic models can be obtained (Table 1).

The model predictions had good agreements with experimental data when cleaning the HA-fouled 0.1 μm PVDF microfiltration membrane using NaCl solution ($R^2 > 0.9897$) (Fig. 20). Meanwhile, the effect of the variation of operating conditions on k_1 and k_2 was also considered and $M_{1+1,REG}$ (k_1 and k_2 took the regression values as a function of different cleaning conditions) gave a better fit to the membrane fouling resistance than that of $M_{1+1,AVE}$ (k_1 and k_2 took average values over different NaCl cleaning conditions).

References

1. H. Ma, C.N. Bowman, R.H. Davis, Membrane fouling reduction by backpulsing and surface modification. J. Membr. Sci. **173**, 191–200 (2000)
2. Z. Wang, C. Zhou, *Foundation of Membrane Separation Technology*, 3rd edn. (Chemistry Industry Press, Beijing, 2019)
3. Z. Wang, J. Ma, C.Y. Tang, K. Kimura, Q. Wang, X. Han, Membrane cleaning in membrane bioreactors: a review. J. Membr Sci **468**, 276–307 (2014)

4. E. Akhondi, F. Zamani, A.W.K. Law et al., Influence of backwashing on the pore size of hollow fiber ultrafiltration membranes. J Membr Sci **521**, 33–42 (2017)
5. A. Grasshoff, Hygienic design: the basis for computer-controlled automation. Food Bioprod Process, Trans Inst Chem Eng Part C **70**, 69–77 (1992)
6. K. Nakanishi, H.G. Kessler, Rinsing behavior of deposited layers formed on membranes in ultrafiltration. J. Food Sci. **50**, 1726–1731 (1985)
7. E.A. Plett, Cleaning of fouled surfaces, in *Proceedings of the Second International Conference on Fouling and Cleaning in Food Processing*, eds. by D. Lund, E. Plett, C. Sandu (Madison, WI, 1985), pp. 286–311
8. E.A. Plett, in *Cleaning and sanitation, Handbook of Food Engineering*. ed. by D.R. Heldman, B. Lund (Marcel Dekker Inc., New York, 1992), pp.719–740
9. Y. Kong, Z. Wang, Y. Ma, L. Hou, W. Yao, Theory investigation on the variation of fouling resistance during water rinsing process of the membrane fouled with sodium alginate. J. Taiwan Inst. Chem. Eng. **66**, 230–238 (2016)
10. R. Krack, Chemical agents and costs in cleaning and disinfection of membrane equipment, in *Fouling and Cleaning in pressure Driven Membrane Processes* (FIL-IDF, Supplement to B-Doc 250, 1994), pp. 210–238
11. M.L. Cabero, F.A. Riera, R.Â lvarez, Rinsing of ultrafiltration ceramic membranes fouledwith whey proteins: effects on cleaning procedures. J. Membr Sci **154**, 239–250 (1999)
12. B. Bansal, R. Al-Ali, R. Mercade-Prieto, X.D. Chena, Rinsing and cleaning of α-lactalbumin fouled MF membranes. Sep Purif Technol **48**, 202–207 (2006)
13. Z. Wang, S.S. Zhao, F. Liu, L. Yang, Y. Song, X. Wang, X. Xi, Influence of operating conditions on cleaning efficiency in sequencing batch reactor (SBR) activated sludge process—water rinsing introduced membrane filtration process. Desalination **259**, 235–242 (2010)
14. P. Matzinos, R. Álvarez, Effect of ionic strength on rinsing and alkaline cleaning of ultrafiltration inorganic membranes fouled with whey proteins. J. Membr. Sci. **208**, 23–30 (2002)
15. M.F. Rabuni, N.M. NikSulaiman, M.K. Aroua, C.Y. Chee, N.A. Hashim, Impact of in situ physical and chemical cleaning on PVDF membrane properties and performances. Chem. Eng. Sci. **122**, 426–435 (2015)
16. R.M. Prieto, X.D. Chen, Caustic-induced gelation of whey deposits in the alkali cleaning of membranes. J. Membr. Sci. **254**, 157–167 (2005)
17. S.S. Popović, M.N. Tekić, M.S. Djuric, Kinetic models for alkali, detergent cleaning of ceramic tubular membrane fouled with whey proteins. J. Food Eng. **94**, 307–315 (2009)
18. Y. Kong, Z. Wang, Y. Ma, H. Wang, B. Khan, Prediction of the instantaneous fouling resistance of sodium alginate during water hydraulic cleaning. Chem. Eng. J. **22**, 121–131 (2017)
19. Y.Y. Guo, Z. Wang Y. Ma, P. Li, G. Hu, A new composite model of the membrane cleaning for predicting the fouling resistance in the hydraulic cleaning process. J. Membr. Sci. **602**, 117957 (2020)
20. E. Tardieua, A. Grasmickb, V. Geaugeyc, J. Manem, Hydrodynamic control of bioparticle deposition in a MBR applied to wastewater treatment. J. Membr. Sci. **147**, 1–12 (1998)
21. P. Saikhwan, R. Mercadé-Prieto, Y.M.J. Chew, S. Gunasekarand, W.R. Paterson, D.I. Wilson, Swelling and dissolution in cleaning of whey protein gels. Food Bioprod Process **88**, 375–383 (2010)
22. R. Mercadé-Prieto, R. J. Falconer, W. R. Paterson, D. I. Wilson, Probing the mechanisms limiting dissolution of whey protein gels during cleaning. Food Bioprod Process **84**, 311–319 (2006)
23. Z. Wang, J. Ma, C.Y. Tang, K. Kimura, Q. Wang, X. Han, Membrane cleaning in membrane bioreactors: a review. J. Membr. Sci. **468**, 276–307 (2014)
24. Z. Wang, Y. Guo, Z. Qiao, A new model considering the morphological changes of cake and gel for the in-situ membrane cleaning process. J. Environ. Chem. Eng. **9**, 106706 (2021)
25. P. Saikhwan, R. Mercade-Prieto, Y.M.J. Chew, S. Gunasekarand, W.R. Patersonb, D.I. Wilson, Swelling and dissolution in cleaning of whey protein gels. Food Bioprod. Process. **88**, 375–383 (2010)

26. R. Mercade-Prieto, R.J. Falconer, W.R. Paterson, D.I. Wilson, Probing the mechanisms limiting dissolution of whey protein gels during cleaning. Food Bioprod. Process. **84**, 311–319 (2006)
27. E. Tardieu, A. Grasmick, V. Geaugey, J. Manem, Hydrodynamic control ofbioparticle deposition in a MBR applied to wastewater treatment. J. Membr. Sci. **147**, 1–12 (1998)
28. Z. Qiao, Z. Wang, Y. Guo, A new model based on the cake removal and the re-deposition mechanism in the rinsing process. J. Membr. Sci. **636**, 119560 (2021)
29. W. Yao, Z. Wang, X. Wang, Detachment mechanism and energy consumption model for the ex-situ rinsing process in membrane bioreactors. J. Membr. Sci. **615**, 118462 (2020)
30. C.J. van Oss, R.J. Good, A.W. Neumann, J.D. Weiser, P. Rosenberg, Comparison between attachment and detachment approaches to the quantitative study of cell adhesion to low-energy solids. Colloid Interface Sci. **59**, 505–515 (1977)
31. C.J. van Oss, Hydrophobicity, and hydrophilicity of biosurfaces. Curr. Opin. Colloid Interface Sci. **2**, 503–512 (1997)
32. J.H. Teng, M.J. Zhang, K.T. Leung, J.R. Chen, H.C. Hong, H.J. Lin, B.Q. Liao, A unified thermodynamic mechanism underlying fouling behaviors of soluble microbial products (SMPs) in a membrane bioreactor. Water Res. **149**, 477–487 (2019)
33. Z. Qiao, Z. Wang, A multi-functional rinsing model based on the cake properties for predicting rinsing efficiency and calculating energy consumption. J. Membr. Sci. **660**, 120849 (2022)
34. L. Zhao, L. Shen, Y. He et al., Influence of membrane surface roughness on interfacial interactions with sludge flocs in a submerged membrane bioreactor. J. Colloid Interface Sci. **446**, 84–90 (2015)
35. C.N. Haas, J. Joffe, Disinfection under dynamic conditions: modification of hom's model for decay. Environ. Sci. Technol. **28**, 1367 (1994)
36. I. Koyuncu, A. Lüttge, M.R. Wiesner, Interferometric observations and kinetic modeling of the chemical cleaning of humic materials deposited on membranes. J. Membr. Sci. **313**, 127–134 (2008)
37. H. Zhang, J. Gao, J. Tao et al., A novel approach to evaluate the permeability of cake layer during cross-flow filtration in the flocculants added membrane bioreactors. Bioresour. Technol. **102**, 11121–11131 (2011)
38. Z.H. Wang, S.S. Zhao, Influence of operating conditions on cleaning efficiency in sequencing batch reactor (SBR) activated sludge process—water rinsing introduced membrane filtration process. Desalination **259**, 235–242 (2010)
39. Z.W. Wang, J.X. Ma et al., Membrane cleaning in membrane bioreactors: a review. J. Membr. Sci. **468**, 276–307 (2014)
40. Z. Wang, Y. Li, P. Song, X. Wan, NaCl cleaning of 0.1μm polyvinylidene fluoride (PVDF) membrane fouled with humic acid (HA). Chem. Eng. Res. Design **132**, 325–337 (2018)

Backwashing

Nomenclature

A	Backwash duration
$A_f(t)$	The effective membrane area (m^2)
A_m	The total area of the membrane (m^2)
B	Backwash strength
C	Filtration duration
C_0	Is the initial concentration of the chemical reagent (wt%)
$d_{m,b}(t)$	The average pore diameter of the fouled membrane at time t (m)
$d_{m,0}$	The pore diameter of the virgin membrane (m)
f	The Fanning friction coefficient
$J(t)$	The permeate flux ($m^3\ m^{-2}\ s^{-1}$)
$J_b(t)$	The backwash flux ($m^3 \cdot m^{-2} \cdot s^{-1}$)
J_0	The initial filtration rate (m/s)
k	The mass transfer coefficient t (m/s)
k_1	The reaction rate constant between the foulant and chemical reagent (s^{-1})
k_2	Removal rates of foulants in the membrane pores, which are (s^{-1})
k_3	Removal rates of foulants on the membrane surface (s^{-1})
$m_c(t)$	The specific mass of the cake layer (kg/m^2)
m_d	The specific mass of matter detached by cleaning operation
$P(t)$	The total trans-membrane pressure (Pa), or the instantaneous trans-membrane pressure (Pa)
$P_b(t)$	The trans-membrane pressure (kPa)
$Q(t)$	The volumetric flowrate at filtration time t ($m^3 \cdot s^{-1}$)
$Q_b(t)$	The backwash flow rate ($m^3 \cdot s^{-1}$)
$R(t)$	The instantaneous fouling resistance (m^{-1})
$R_b(t)$	The total resistance in the backwash process (m^{-1})
R_c	The cake resistance (m^{-1})
R_{rvf}	The reversible fouling resistance (m^{-1})

Z. Wang (ed.), *Modelling of the Microfiltration Process*,
https://doi.org/10.1007/978-981-99-0471-6_5

Re	Reynolds number
R_{rdf}	The residual fouling resistance (m^{-1})
$R_f(t)$	The resistance value of foulants (m^{-1}) at time t
$R_t(t)$	The total resistance (m^{-1})
$R_{pore}(t)$	The instantaneous fouling resistance of the foulants in the membrane pores (m^{-1})
$R_{surface}(t)$	The instantaneous fouling resistance of the foulants on the membrane surface (m^{-1})
R_{if}	The internal fouling resistance (m^{-1})
$R_{irr,N=1}$	The irreversible resistance in the first cycle, which is a function of backwash flux and duration (m^{-1})
R_m	The virgin membrane resistance (m^{-1})
R_{ef}	External fouling resistance (m^{-1})
R_0	The intrinsic membrane resistance (m^{-1})
TMP_i^{final}	The final TMP value of filtration cycle i
$TMP_{i+1}^{initial}$	The beginning TMP value of filtration cycle i+1
u	Flow velocity in membrane pore (m/s)
$V_b(t)$	The accumulated backwash solution volume (m^3)

Greek Symbols

α	The specific cake resistance (m/kg)
μ	Viscosity of permeate (Pa·s)
ρ	Density of filtrate (kg/m^3)
σ	The recovery of $A_b(t)$ by using unit backwash solution volume (m^{-1})
τ	The shear force in the membrane pore (s^{-1})

1 Introduction

Backwashing refers to the backwashing operation of applying backwashing pressure on the permeate side of the membrane to reverse osmosis membrane holes with permeate or a mixture of permeate and air (Fig. 1). Compared with the water flushing process, backwashing requires high membrane performance, which can not only restore membrane performance (flux or trans-membrane pressure), but also obtain higher flushing efficiency (Fig. 2) [1]. In essence, backwashing mainly uses the shear stress generated by the fluid movement in the membrane pore to remove the foulants in the membrane pore [1]. The common operation mode of backwashing is constant pressure and constant flow mode (Fig. 3).

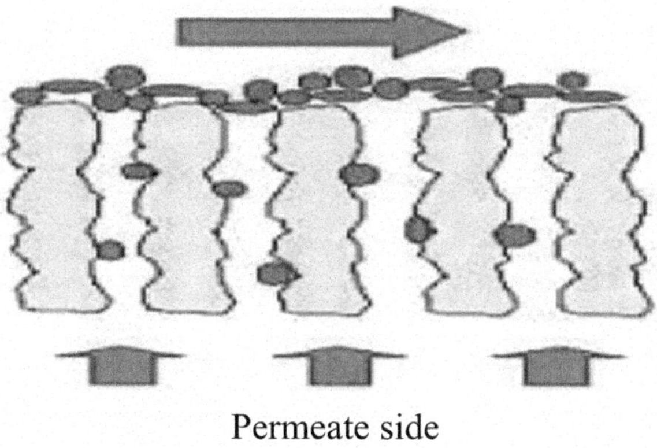

Fig. 1 Schematic diagram of the backwashing process

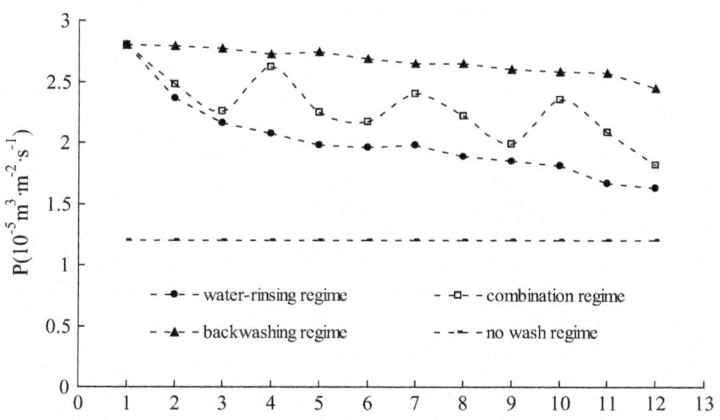

Fig. 2 The variety of the washing efficiency for three different washing regimes as a function of washing cycles to comparing with no wash regime [2]

The backwashing effect is affected by backwashing times, flow rate, operating pressure, backwashing solution temperature, backwashing flux, backwashing duration, backwash interval backwash strength, detergent volume, filtration condition, membrane properties, operation conditions, backwash interval, feed water property, and so on [3–6] (Fig. 4). To avoid membrane damage and water outflow deterioration, the backwashing operation should be performed under low pressure.

The backwashing operation has been intensively studied by vast scientific researchers. Relative ratio of the non-removable resistance to the total fouling resistance (RRNR) can be reduced by increasing backwashing duration, times, operating pressure, and backwashing solution temperature. For example, Yang et al. [2] pointed out that the increase in water backwashing duration, times, operating pressure, and water temperature can reduce the RRNR of a 0.1 μm PES membrane fouled by

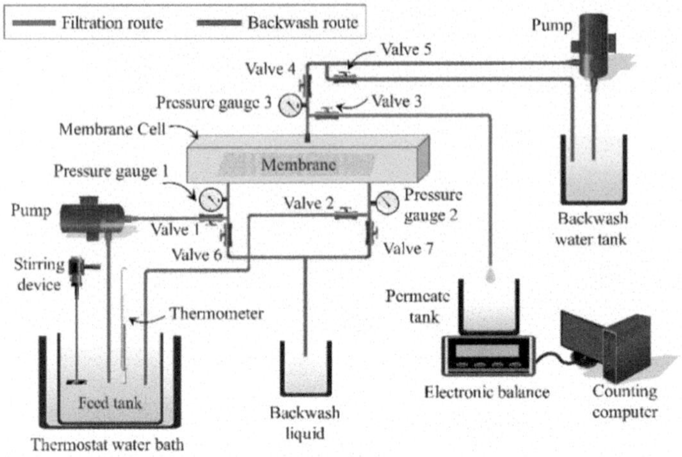

(a) Constant pressure mode

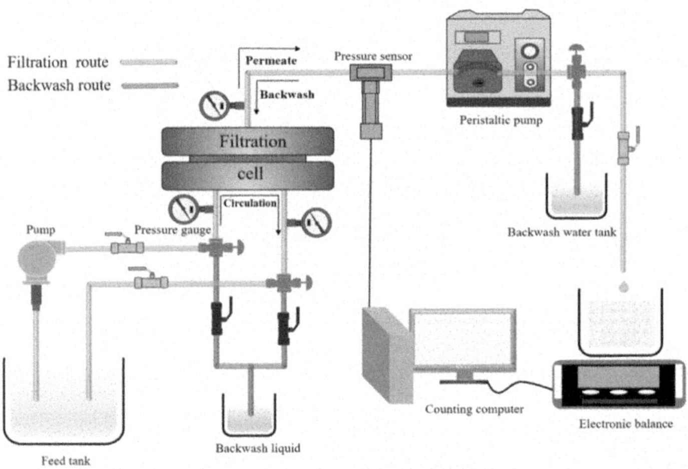

(b) Constant flowrate mode

Fig. 3 System of constant pressure filtration and backwashing (**a**) and system of constant flowrate filtration and backwashing (**b**)

activated sludge suspension. Hwang et al. [3] indicated that the RRNR for a carbon polymer membrane fouled by humic acid decreased by 31% when TMP was increased from 98.4 to 196.8 kPa. (2) Reducing the concentration of sodium and calcium ions in the backwash solution can also reduce RRNR. Li et al. [4] found that RRNR decreased by 4.3% in backwash operation when calcium ion concentration in water decreased from 3 to 0 mmol/L for a PES ultrafiltration membrane fouled by natural organics. (3) RRNR values of different fouling systems in the backwashing operation differ greatly: 7.8% for a 0.1 μm PES membrane fouled by the activated sludge

Fig. 4 The factors influencing on the backwashing efficiency

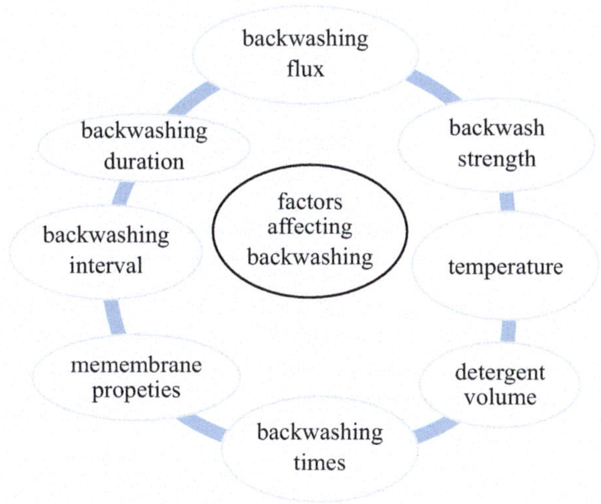

suspension [3], 17% for the PVDF UF membrane fouled by sodium alginate [7]; 16% for a PES ultrafiltration membrane fouling by humic acid is [8] and 28.1% for a PES ultrafiltration membrane fouled by natural organic matter (NOM) [4]. It was found that the backwashing operation with low frequency and long duration (45 s per 600 s of filtration) was better than the cleaning operation with high frequency and short duration (15 s per 200 s of filtration) [9]. In addition, although the operation with high frequency and long backwashing duration can remove more foulants, it is necessary to optimize the parameters of the backwashing operation in consideration of energy and detergent consumption. By detecting the change in operating pressure difference during filtration, the automatic control system can optimize the backwashing duration [10].

The magnitude of the shear force in the membrane hole can be obtained from the following equation [11]:

$$\tau = \frac{f}{2}\rho u^2 \tag{1}$$

where τ is the shear force (s^{-1}) in the membrane pore, f is the Fanning friction coefficient, ρ is the density (kg/m^3) of the cleaning agent, and u is the flow velocity of the cleaning agent in the membrane pore (m/s).

The Fanning friction coefficient can be expressed as

$$f = \frac{16}{\mathrm{Re}} \tag{2}$$

where Re is Reynolds numbers.

From Eqs. (1) and (2):

$$\tau = 8\frac{u \cdot \mu}{d_{m,b}(t)} \tag{3}$$

where $d_{m,b}(t)$ is the average pore diameter of the fouled membrane at time t (m), which can be calculated by the following equation [12]:

$$d_{m,b}(t) = \left(d_{m,0} \cdot \frac{R_m}{R_f(t)}\right)^{0.5} \tag{4}$$

where $d_{m,0}$ is the pore diameter of the virgin membrane (m), R_m is the inherent resistance of the membrane (m^{-1}), and $R_f(t)$ is the resistance value of foulants (m^{-1}) at time t.

The mass transfer coefficient (k) in the backwashing process can be derived from Deissler's theorem [13]:

$$k = 0.023\left(\frac{\rho}{\mu}\right)^{0.625} u^{0.875} \frac{D^{0.75}}{d_{m,b}^{0.125}(t)} \tag{5}$$

where k is the mass transfer coefficient t (m/s), ρ is the density of the cleaning agent (kg/m^3), μ is the viscosity of the cleaning agent (Pa·s), u is the liquid flow rate in the membrane pore (m/s), and D is the diffusion coefficient (m^2/s).

The variations of shearing stress TMP and the corresponding mass transfer coefficient with the increasing of TMP in the dead-end backwashing process are shown in Fig. 5 [2].

For a tertiary submerged membrane bioreactor (MBR) [14]:

$$\mathrm{TMP}(t) = J\mu R_T = J\mu(R_m + R_{\mathrm{rdf}} + R_{\mathrm{rvf}}) \tag{6}$$

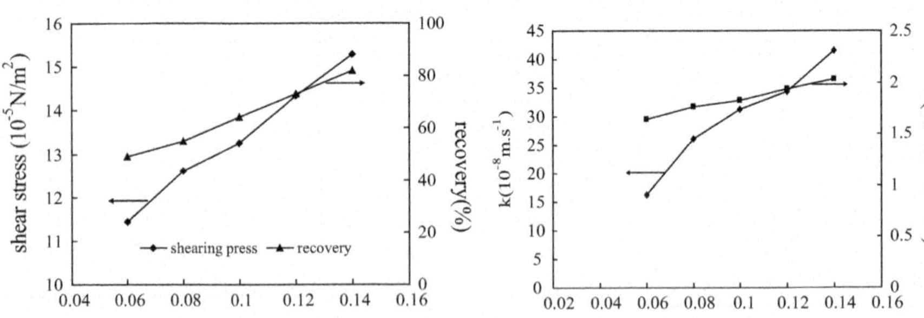

(a) The variation of shearing stress with TMP

(b) the variation of mass-transfer coefficient with TMP

Fig. 5 The variations of shearing stress TMP and the corresponding mass transfer coefficient with TMP in the dead-end backwashing process [2]

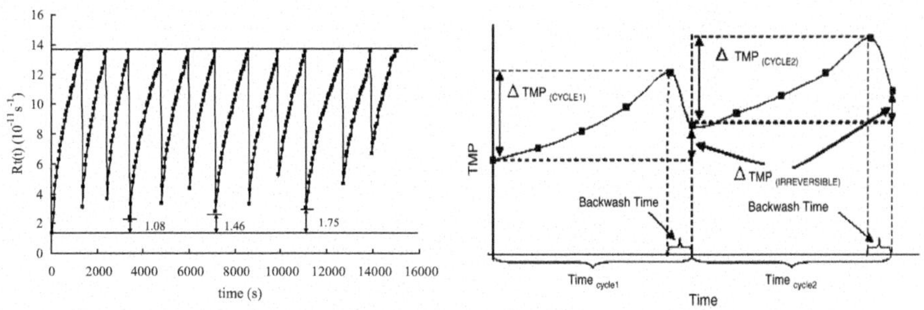

(a) The variation of the irreversible resistance (R_{irr}) in constant pressure mode [1]

(b) The variation of irreversible ΔTMP in constant flow rate mode [19]

Fig. 6 Graphical presentation of a membrane system operating with an intermittent periodic permeate production cycle and a periodic backwash cycle

where μ is the permeate viscosity (Pa·s), J is the permeate flux (m/s), R_T is the total hydraulic resistance (m^{-1}), R_m is the clean membrane resistance (m^{-1}), R_{rdf} is the residual fouling resistance (m^{-1}), and R_{rvf} is the reversible fouling resistance (m^{-1}).

Residual fouling is that which is not removed by backwashing (Fig. 6a). The resistance formed by residual fouling can be separated into internal fouling resistance (R_{if}), related to internal membrane pore blocking [15] and gel or compacted particle layer that is not detached during backwashing [16], and external fouling resistance (R_{ef}) linked to fouling layer recompression [17]. Therefore, R_{rdf} can be expressed as

$$R_{rdf} = R_{if} + R_{ef} \tag{7}$$

For a submerged membrane filtration by using a ceramic tubular membrane with a pore size of 5.6 μm, the effects of backwash flux and duration on the irreversible resistance (R_{irr}) in the constant pressure filtration with a periodic backwash were systemically studied [18]. The relationship between irreversible resistance (R_{irr}) and cycle number (N) can be expressed in a power-type empirical equation:

$$R_{irr} = R_{irr,N=1} N^{0.75} \tag{8}$$

where $R_{irr,N=1}$ is the irreversible resistance in the first cycle, which is a function of backwash flux and duration.

The increase of the irreversible resistance in each backwashing cycle also can be expressed as the increase of the irreversible TMP ($\Delta TMP_{(IRREVERSIBLE)}$) in each cycle for constant flowrate mode (Fig. 6b) [19].

2 Water Rinsing Model

2.1 Ye's Model [20]

The operation parameters such as filtration duration, backwash duration, backwash strength and the aid of air scouring during backwash were investigated.

$$\text{Reversible fouling }(\%) = \frac{\text{TMP}_i^{\text{final}} - \text{TMP}_{i+2}^{\text{initial}}}{\Delta\text{TMP}_i} \tag{9}$$

where $\text{TMP}_i^{\text{final}}$ is the final TMP value of filtration cycle *i*, $\text{TMP}_{i+1}^{\text{initial}}$ is the beginning TMP value of filtration cycle *i*+1 and ΔTMP_i indicates the TMP increased in the filtration cycle *i*.

Based on the regression analysis, the relationship between the objective function (ΔTMP_f (it represents the total TMP increase between the beginning and the end of the filtration), $d(\Delta\text{TMP}/dt_i)$ (it represents the TMP increase rate or the fouling rate in the filtration cycle *i*) and reversibility and process variables (*A* (backwash duration), *B* (backwash strength) and *C* (filtration duration)) was obtained as follows:

$$\begin{aligned}\Delta\text{TMP}_f &= 14.3 + \frac{9.95}{2}\left(\frac{C-1}{0.5}\right) - \frac{3.9}{2}\left(\frac{A-50}{40}\right) - \frac{3.15}{2}\left(\frac{B-1.25}{0.75}\right)\\ &\quad - \frac{2.74}{2}\left(\frac{A-50}{40}\right)\left(\frac{B-1.25}{0.75}\right)\end{aligned} \tag{10}$$

$$\frac{d\text{TMP}}{dt} = 3.0 + \frac{0.85}{2}\left(\frac{C-1}{0.5}\right) - \frac{0.53}{2}\left(\frac{B-1.25}{0.75}\right)\left(\frac{C-1}{0.5}\right) \tag{11}$$

$$\begin{aligned}\text{reversibility} &= 0.86 - \frac{0.076}{2}\left(\frac{A-50}{40}\right)\left(\frac{B-1.25}{0.75}\right) + \frac{0.071}{2}\left(\frac{A-50}{40}\right)\\ &\quad + \frac{0.06}{2}\left(\frac{B-1.25}{0.75}\right) - \frac{0.057}{2}\left(\frac{C-1}{0.5}\right)\end{aligned} \tag{12}$$

A (backwash duration), *B* (backwash strength), and *C* (filtration duration) were all the influential factors of ΔTMP_f, $\frac{d\text{TMP}}{dt}$ and reversibility. Meanwhile, no important interaction of these three factors on the ΔTMP_f was observed and the sequence of influencing degree for ΔTMP_f was $C > A > B$. In terms of $\frac{d\text{TMP}}{dt}$, *C* is the most important effect, followed by the interaction of *B* and *C*. The significant effects are the interaction of *A* and *B* followed by *A* for reversibility.

2.2 Charfi's Model [21]

In a constant flowrate dead-end process, the TMP variation with time can be expressed as

$$\text{TMP} = J\mu(R_0 + R_C) \tag{13}$$

where J is the permeate flux (m/s), μ is the permeate viscosity (Pa·s), R_0 is the intrinsic membrane resistance (m^{-1}) and R_c is the cake resistance (m^{-1}).

$$R_C = \alpha \cdot m_c \tag{14}$$

where α is the specific resistance of the cake (m/kg) and m_c is the specific mass of the cake layer (kg/m^2).

According to the mass balance:

$$\frac{dm_c}{dt} = \frac{dm_a}{dt} - \frac{dm_d}{dt} \tag{15}$$

where m_c is the specific mass of matter dragged by convective forces and attached to the membrane surface and m_d is the specific mass of matter detached by cleaning operation.

Suppose that the specific mass attached to the membrane is proportional to the permeate flux and the total foulants concentration:

$$\frac{dm_a}{dt} = J \cdot C \tag{16}$$

where J is the permeate flux ($m^3\ m^{-2}\ s^{-1}$) and C is the foulants concentration (kg/m^3).

Assume that the specific mass removed by the backwashing operation is proportional to the attached specific mass and the accumulated specific mass:

$$\frac{dm_d}{dt} = \gamma \cdot \frac{dm_a}{dt} \cdot m_c \tag{17}$$

Suppose that in the periodic cleaning stage, the reduction of the specific mass of the cake can be expressed by a first-order differential equation:

$$\frac{dm_c}{dt} = -\sigma \cdot m_c \tag{18}$$

After introducing Eqs. (16)–(18) into Eq. (15):

$$m_c = \frac{J \cdot C}{\gamma \cdot J \cdot C - \sigma} \tag{19}$$

By introducing Eq. (19) into Eq. (14):

$$R_C = \alpha \cdot m_c = \alpha \cdot \frac{J \cdot C}{\gamma \cdot J \cdot C - \sigma} \tag{20}$$

Assuming that the specific resistance of the cake decreases in the cleaning stage and a first-order differential equation can express it

$$\frac{d\alpha}{dt} = -k_1 \cdot \alpha \tag{21}$$

From Eq. (21):

$$\alpha = \alpha_0 e^{-k_1 t} \tag{22}$$

By introducing Eq. (22) into Eq. (20):

$$R_C = \alpha_0 e^{-k_1 t} \cdot \frac{JC}{\gamma JC - \alpha_0 e^{-k_2 t}} \tag{23}$$

Therefore, a model simulates the TMP variation with time in the constant flowrate dead-end process can be obtained:

$$\text{TMP} = J\mu(R_0 + R_C) = J\mu(R_0 + \alpha_0 e^{-k_2 t} \cdot \frac{JC}{\gamma JC - \alpha_0 e^{-k_2 t}}) \tag{24}$$

The results showed that the model predictions agreed with experimental data (Fig. 7).

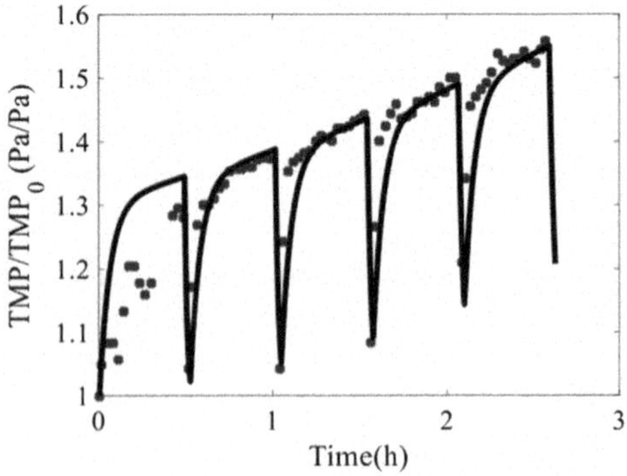

Fig. 7 Comparison of the model productions and experimental data [21]

2.3 Gao's Model [22]

The permeate flux ($J_f(t)$) in the filtration process can be expressed as [23]

$$J_f(t) = \frac{p_f(t)}{R_f(t)\mu} = \frac{Q_f(t)}{A_f(t)} \tag{25}$$

where $J_f(t)$ ($m^3 \cdot m^{-2} \cdot s^{-1}$) is the permeate flux in the filtration process, $R_f(t)$ (m^{-1}) is the total resistance, $P_f(t)$ (kPa) is the trans-membrane pressure, μ (kPa·s^{-1}) is the permeate viscosity, $Q_f(t)$ ($m^3 \cdot s^{-1}$) is the permeate flowrate and $A_f(t)$ (m^2) is the effective membrane area.

For a constant flowrate process:

$$Q_f(t) = \frac{P_f(t)}{R_f(t)\mu} A_f(t) \tag{26}$$

Because the membrane pores are blocked by the activated sludge particles and the cake is formed on the membrane surface in the constant flowrate process, $P_f(t)$ and $R_f(t)$ will increase with time, while $A_f(t)$ will decrease with time.

In the backwash process [24]:

$$Q_b(t) = \frac{p_b(t) A_b(t)}{R_b(t)\mu} \tag{27}$$

where $Q_b(t)$ is the backwash flow rate ($m^3 \cdot s^{-1}$), $P_b(t)$ is the trans-membrane pressure (kPa), $A_b(t)$ is the effective membrane area (m^2) and $R_b(t)$ is the total resistance in the backwash process (m^{-1}).

At the backwash time of 0:

$$Q_b(0) = \frac{P_b(0) A_b(0)}{R_b(0)\mu} \tag{28}$$

Therefore, the trans-membrane pressure ($P_b(t)$) in the constant flowrate backwash process [23]:

$$P_b(t) = \frac{A_b(0) R_b(t)}{A_b(t) R_b(0)} P_b(0) \tag{29}$$

The effective membrane area ($A_b(t)$) in the backwash process can be considered as a function of the accumulated backwash solution volume ($V_b(t)$):

$$A_b(t) = A_b(0) + \sigma V_b(t) \tag{30}$$

where $V_b(t)$ (m^3) is the accumulated backwash solution volume and σ (m^{-1}) is the recovery of $A_b(t)$ using unit backwash solution volume.

According to the report [25]:

$$V_b(t) = Q_b t = A_b(t) J_b(t) t \tag{31}$$

where $J_b(t)$ ($m^3 \cdot m^{-2} \cdot s^{-1}$) is the backwash flux.

After substitution from Eq. (31) into Eq. (30) gives

$$A_b(t) = A_b(0) + \sigma Q_b t \tag{32}$$

Therefore

$$\frac{dA_b(t)}{dt} = \sigma Q_b \tag{33}$$

In the constant flowrate (Q_b) backwash process, $J_b(t)$ decreased according to $J_b(t) = Q_b / A_b(t)$ [26]. However, $A_b(t)$ increased sharply in the early stage and then reached a plateau instead of a linear increase with time. Therefore, a coefficient k was introduced to $A_b(t)$ with backwash time.

$$\frac{dA_b(t)}{dt} = \sigma Q_b - k(A_b(t) - A_b(0)) \tag{34}$$

where k (s^{-1}) is a coefficient.

Rearrangement led to

$$\int_{A_b(0)}^{A_b(t)} \frac{1}{\sigma Q_b - k(A_b(t) - A_b(0))} dA_b(t) = \int_0^t t dt \tag{35}$$

After integration:

$$\frac{A_b(t)}{A_m} = \frac{A_b(0)}{A_m} + \frac{\sigma Q_b}{A_m k}(1 - \exp(-kt)) \tag{36}$$

where A_m (m^2) is the total area of the membrane.

The theoretical physical limitation of $A_b(t)/A_m$ was in the range of $A_b(0)/A_m$ to 1. However, $A_b(t)/A_m$ will keep a steady, the effective membrane area ratio (A_{steady}/A_m) when the process of the open pores is completed instead of reaching A_m in the actual backwash process (Fig. 8).

Moreover, the total resistance is constantly decreased in the backwash process (Fig. 8) and the variation of the total resistance ($R_b(t)$) with backwash time can be described as [27, 28]

$$R_b(t) = R_{irr}(t) + (R_b(0) - R_{irr}(t)) \exp(-k_1 t) \tag{37}$$

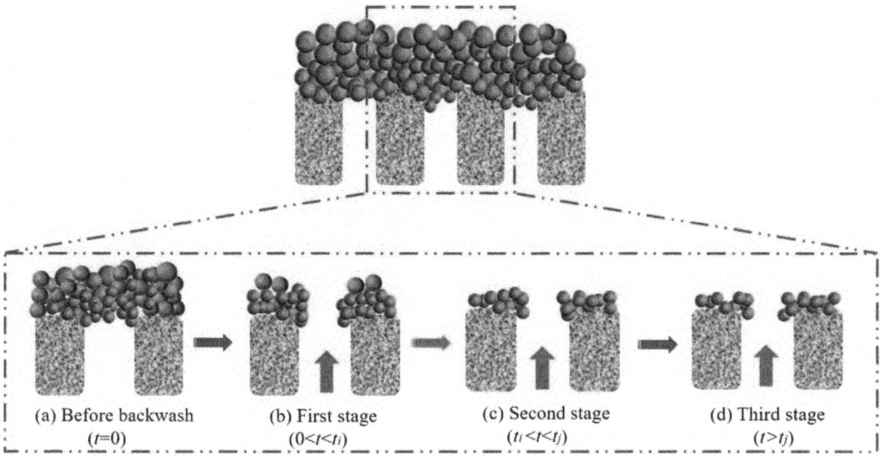

Fig. 8 The backwash mechanism

After substituting Eqs. (36) and (37) into Eq. (29), the trams-membrane pressure ($P_b(t)$) as a function of the backwash time can be obtained:

$$P_b(t) = P_b(0)\left(\frac{(R_b(0) - R_{\text{irr}}(t))\exp(-k_1 t)}{R_b(0)\left[1 + \frac{\sigma Q_b}{k A_b(0)}(1 - \exp(-kt))\right]} + \frac{R_{\text{irr}}(t)}{R_b(0)}\right) \tag{38}$$

The predictions of the proposed model had good agreements with experimental data ($R^2 > 0.983$) at different filtration conditions (cross-flow velocities (7.5, 15, and 22.5 cm/s), filtration fluxes (50, 70, 105 LMH), and feed concentrations (2.29, 3.11, 5.46 g/L)) using a 0.1 μm PAN membrane fouled with activated sludge suspension (Fig. 9) [22].

2.4 *Qiao's Model [29]*

The irreversible foulants on the membrane can react with the chemical reagent and further weaken the attractive force between the foulants and foulants/membrane [30] (Fig. 10a). Finally, it can be carried away from the membrane by the drag force produced by fluid movement [2].

Assume that *A*-pore, *B*-pore, and *C*-pore represent the virgin irreversible foulants, the foulant that reacts with the chemical reagent, and the foulant is removed by the CEB solution in the membrane pores, respectively. *A*-surface, *B*-surface, and *C*-surface represent the virgin irreversible foulants, the foulant that reacts with the chemical reagent, and the foulant removed by the CEB solution on the membrane surface, respectively, (Fig. 10b). The corresponding instantaneous fouling resistances of the

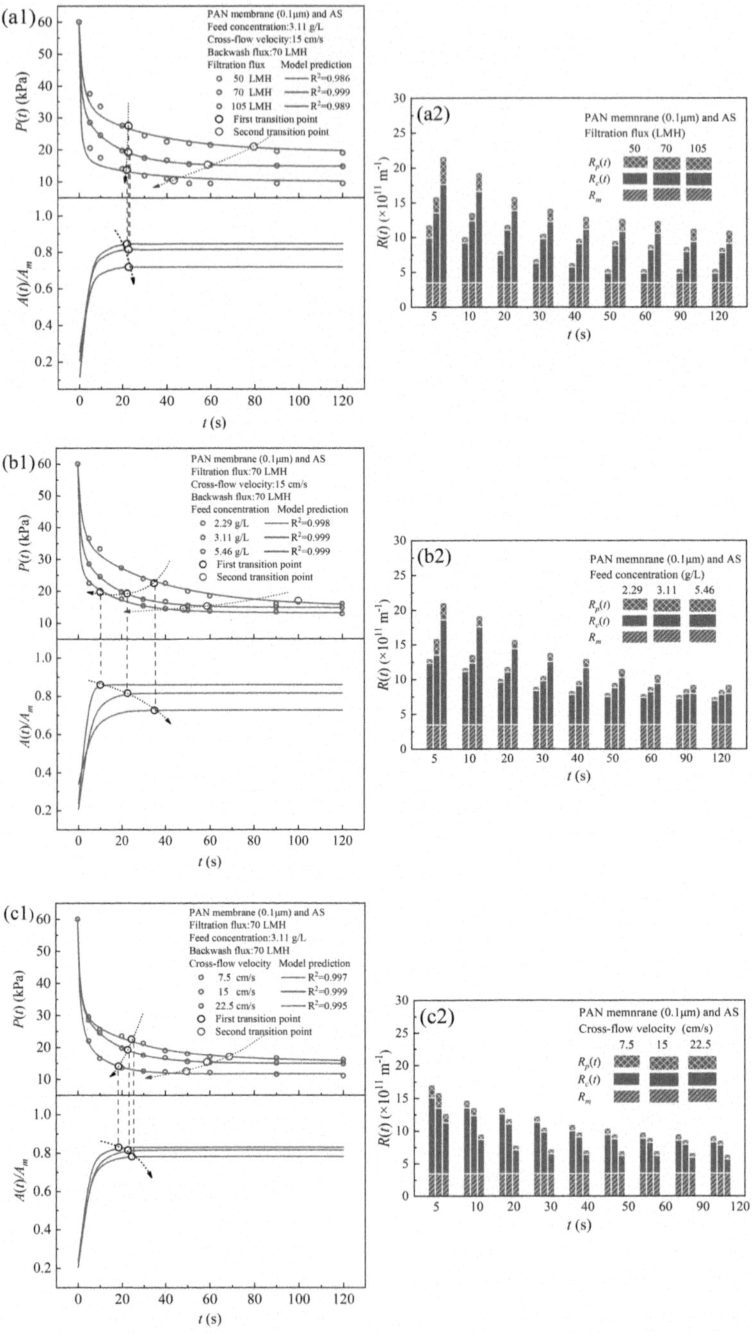

Fig. 9 The variation of the trans-membrane pressure ($P(t)$), the available membrane area ratio ($A(t)/A_m$) and the resistance ($R(t)$) with the backwash duration under the different filtration fluxes ((**a1**) and (**a2**)), feed concentration ((**b1**) and (**b2**)) and cross-flow velocity ((**c1**) and (**c2**))

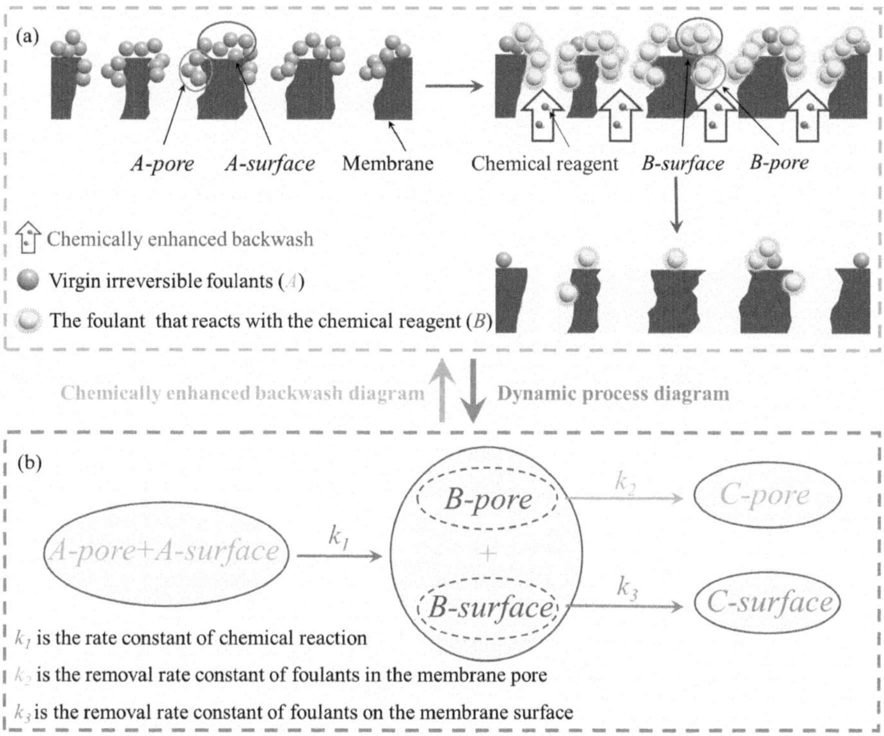

Fig. 10 The chemically enhanced backwash (CEB) mechanism of the fouled membrane and the schematic diagram of the dynamic variation of irreversible foulants in the CEB process

A-pore, *A*-surface, *B*-pore, *B*-surface, *C*-pore, and *C*-surface in the CEB process can be expressed as $R_{A\text{-pore}}(t)$, $R_{A\text{-surface}}(t)$, $R_{B\text{-pore}}(t)$, $R_{B\text{-surface}}(t)$, $R_{C\text{-pore}}(t)$, and $R_{C\text{-surface}}(t)$ (m^{-1}), respectively. k_1 is the reaction rate constant between the foulant and chemical reagent (s^{-1}). In addition, the foulants in the membrane pores and on the membrane surface suffer different drag forces [31] and also results in different removal rates of foulants in the membrane pores and on the membrane surface, which are k_2 (s^{-1}) and k_3 (s^{-1}), respectively. Therefore, the instantaneous fouling resistance ($R(t)$, m^{-1}) can be expressed as follows:

$$R(t) = R_{\text{pore}}(t) + R_{\text{surface}}(t) = R_{A\text{-pore}}(t) + R_{B\text{-pore}}(t) + R_{A\text{-surface}}(t) + R_{B\text{-surface}}(t) \tag{39}$$

where $R_{\text{pore}}(t)$ is the instantaneous fouling resistance of the foulants in the membrane pores (m^{-1}), $R_{\text{surface}}(t)$ is the instantaneous fouling resistance of the foulants on the membrane surface (m^{-1}), and t is the CEB time (s).

According to Fig. 10, the corresponding kinetic equations can be obtained as follows:

$$\frac{dR_{A\text{-pore}}(t)}{dt} = -k_1 C_0^n R_{A\text{-pore}}(t) \tag{40}$$

$$\frac{dR_{B\text{-pore}}(t)}{dt} = k_1 C_0^n R_{A\text{-pore}}(t) - k_2 R_{B\text{-pore}}(t) \tag{41}$$

$$\frac{dR_{C\text{-pore}}(t)}{dt} = k_2 R_{B\text{-pore}}(t) \tag{42}$$

where C_0 is the initial concentration of the chemical reagent (wt%) (C_0 is constant in the CEB process because the CEB time is short and the CEB solution is in a moving state) and n is the empirical constant.

Similarly, the corresponding kinetic equations for variation in foulants on the membrane surface can also be expressed as follows:

$$\frac{dR_{A\text{-surface}}(t)}{dt} = -k_1 C_0^n R_{A\text{-surface}}(t) \tag{43}$$

$$\frac{dR_{B\text{-surface}}(t)}{dt} = k_1 C_0^n R_{A\text{-surface}}(t) - k_3 R_{B\text{-surface}}(t) \tag{44}$$

$$\frac{dR_{C\text{-surface}}(t)}{dt} = k_3 R_{B\text{-surface}}(t) \tag{45}$$

According to Darcy's law, Eqs. (40), (41), (43), and (44) are substituted into Eq. (39), and the instantaneous trans-membrane pressure ($p(t)$, Pa) can be expressed as

$$p(t) = \mu J_0\left(R_m + R(0)e^{-k_1 C_0^n t} + \frac{k_1 C_0^n R_{A\text{-pore}}(0)}{(k_2 - k_1 C_0^n)}(e^{-k_1 C_0^n t} - e^{-k_2 t}) + \frac{k_1 C_0^n R_{A\text{-surface}}(0)}{(k_3 - k_1 C_0^n)}(e^{-k_1 C_0^n t} - e^{-k_3 t})\right) \tag{46}$$

where μ is the permeate viscosity (Pa·s), J_0 is the constant permeate flux of the membrane ($m^3 \cdot m^{-2} \cdot s^{-1}$), and R_m is the virgin membrane resistance (m^{-1}).

The corresponding $R_{\text{pore}}(t)$ and $R_{\text{surface}}(t)$ gradually decrease in the CEB process. Similar to the definition of the boundary layer theory [32, 33], $R_{\text{pore}}(t)$ decreases to $0.01R_{\text{pore}}(0)$ at the CEB time t_i ($R_{\text{pore}}(t_i) = 0.01R_{\text{pore}}(0)$), which indicates that the foulants in the membrane pores have been completely removed. Similarly, the removal of foulants on the membrane surface is completed when $R_{\text{surface}}(t)$ decreases to $0.01R_{\text{surface}}(0)$ at the CEB time t_j ($R_{\text{surface}}(t_j) = 0.01R_{\text{surface}}(0)$). The corresponding t_i and t_j values can be determined using Eqs. (47) and (48), respectively.

$$\begin{aligned} R_{\text{pore}}(t_i) &= R_{A\text{-pore}}(0)e^{-k_1 C_0^n t_i} + \frac{k_1 C_0^n R_{A\text{-pore}}(0)}{(k_2 - k_1 C_0^n)}(e^{-k_1 C_0^n t_i} - e^{-k_2 t_i}) \\ &= 0.01 R_{\text{pore}}(0) \end{aligned} \tag{47}$$

$$R_{\text{surface}}(t_j) = R_{A\text{-surface}}(0)e^{-k_1 C_0^n t_j} + \frac{k_1 C_0^n R_{A\text{-surface}}(0)}{(k_3 - k_1 C_0^n)}(e^{-k_1 C_0^n t_j} - e^{-k_3 t_j})$$
$$= 0.01 R_{\text{surface}}(0) \quad (48)$$

The predictions of the proposed model had good agreements with experimental data ($R^2 > 0.991$) for the instantaneous trans-membrane pressure ($p(t)$) in the NaOCl-assisted backwash process of 0.1 μm PAN membranes fouled with yeast suspension decreased gradually with increasing the chemically enhanced backwash (CEB) time (Fig. 11). As a result, the oxidation of NaOCl could enhance the hydrophilicity of the foulant and break up large organic particles, further weakening the attractive force between the foulant and foulant/membrane [34]. These irreversible foulants are then removed by the drag force in the NaOCl-assisted backwash process.

The model predictions were in good agreement with the experimental data ($R^2 > 0.970$) for a series of multiple filtrations and cleaning experiments for 0.1 μm PAN membranes fouled with yeast suspension were performed under different CEB conditions (Fig. 12) [29].

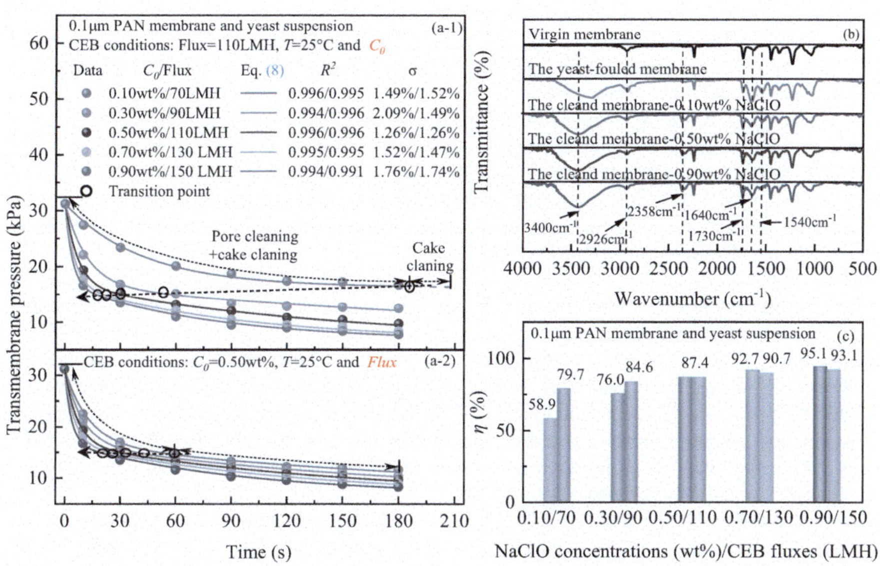

Fig. 11 Influence of cleaning conditions on cleaning effect. **a** The effect of CEB conditions (**a-1** NaOCl concentration and **a-2** CEB flux) on $p(t)$, **b** FTIR spectra of the virgin membrane, yeast-fouled membrane, and CEB-cleaned membrane, and **c** the effect of CEB conditions on the cleaning efficiency (η) [29]

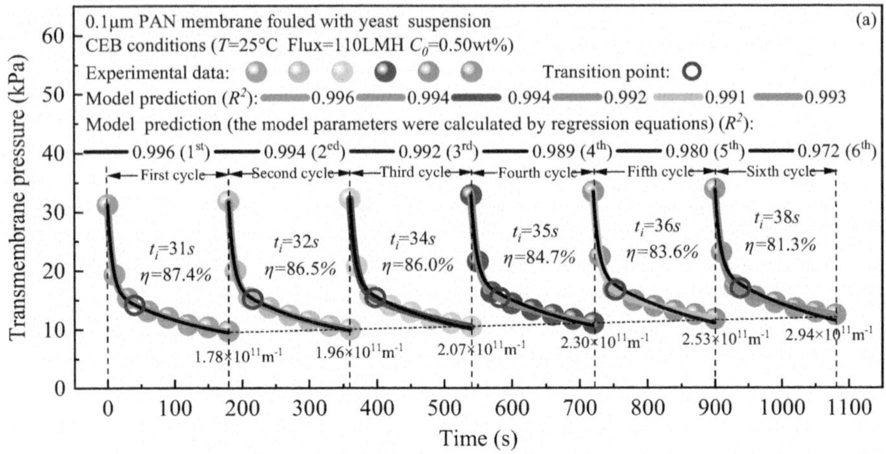

Fig. 12 Model verification

References

1. H. Chang, H. Liang, F. Qu, J. Ma, N. Ren, G. Li, Towards a better hydraulic cleaning strategy for ultrafiltration membrane fouling by humic acid: effect of backwash water composition. J. Environ. Sci. **4**(3), 177–186 (2016)
2. L.Y. Yang, Z. Wang, Y.J. Sun et al., Influence of various operating conditions on cleaning efficiency in sequencing batch reactor (SBR) activated sludge process. Part II: backwash and water rinsing introduced membrane filtration process. Desalination **272**, 76–84 (2011)
3. L.L. Hwang, M.Y. Wey, J.C. Chen, Effects of membrane compositions and operating conditions on the filtration and backwashing performance of the activated carbon polymer composite membranes. Desalination **352**, 181–189 (2014)
4. S. Li, S.G.J. Heijman, J.Q.J.C. Verberk, A.R.D. Verliefde, G.L. Amya, J.C. van Dijk, Removal of different fractions of NOM foulants during demineralized water backwashing. Sep. Purif. Technol. **98**, 186–192 (2012)
5. Y. Gao, J. Qin, Z. Wang, S.W. Østerhus, Backpulsing technology applied in MF and UF processes for membrane fouling mitigation: a review. J. Membr. Sci. **587**, 117136 (2019)
6. H. Chang, H. Liang, F. Qu, B. Liu, H. Yu, X. Du, G. Li, Hydraulic backwashing for low-pressure membranes in drinking water treatment: a review. J. Membr. Sci. **540**, 362–380 (2017)
7. A. Resosudarmo, Y. Ye, P. Le-Clech, V. Chen, Analysis of UF membrane fouling mechanisms caused by organic interactions in seawater. Water Res. **47**, 911–921 (2013)
8. K. Katsoufidou, S.G. Yiantsios, A.J. Karabelas, A study of ultrafiltration membrane fouling by humic acids and flux recovery by backwashing: experiments and modeling. J. Membr. Sci. **266**, 40–50 (2005)
9. T. Jiang, M.D. Kennedy, B.F. Guinzbourg, P.A. Vanrolleghem, J.C. Schippers, Optimising the operation of a MBR pilot plant by quantitative analysis of the membrane fouling mechanism. Water Sci. Technol. **51**, 19–25 (2005)
10. P.J. Smith, S. Vigneswaran, H.H. Ngo, R. Ben-Aim, H. Nguyen, Design of a generic control system for optimising back flush durations in a submerged membrane hybrid reactor. J. Membr. Sci. **255**, 99–106 (2005)
11. J.M. Coulson, J.E. Richardson, Chem. Eng. **1**, 35–70 (1977)
12. K.-J. Hwang, P.-S. Huang, Cross-flow microfiltration of dilute macromo-lecular suspension. Sep. Purif. Technol. **68**, 328–334 (2009)

13. Permeate flux prediction in apple juice concentration by reverse osmosis. J. Membr. Sci. **127**(1), 25–34 (1997)
14. L. Vera, E. González, O. Díaz, S. Delgado, Application of a backwashing strategy based on transmembrane pressure set-point in a tertiary submerged membrane bioreactor. J. Membr. Sci. **470**, 504–512 (2014)
15. T. Jiang, M.D. Kennedy, W.G.J. Van der Meer, P.A. Vanrolleghem, J.C. Schippers, The role of blocking and cake filtration in MBR fouling. Desalination **157**, 335–343 (2003)
16. E.J. McAdam, S.J. Judd, Optimisation of dead-end filtration conditions for an immersed anoxic membrane bioreactor. J. Membr. Sci. **325**, 940–946 (2008)
17. V. Diez, D. Ezquerra, J.L. Cabezas, A. García, C. Ramos, A modified method for evaluation of critical flux, fouling rate and in situ determination of resistance and compressibility in MBR under different fouling conditions. J. Membr. Sci. **453**, 1–11 (2014)
18. K.-J. Hwang, C.-S. Chan, K.-L. Tung, Effect of backwash on the performance of submerged membrane filtration. J. Membr. Sci. **330**, 349–356 (2009)
19. P.J. Smith, S. Vigneswaran, H.H. Ngo, R. Ben-Aim, H. Nguyen, A new approach to backwash initiation in membrane systems. J. Membr. Sci. **278**, 381–389 (2006)
20. Y. Ye, L.N. Sim, B. Herulah, V. Chen, A.G. Fane, Effects of operating conditions on submerged hollow fiber membrane systems used as pre-treatment for seawater reverse osmosis. J. Membr. Sci. **365**, 78–88 (2010)
21. A. Charfi, H. Jang, J. Kim, Modelling tool to assess membrane regeneration by periodical hydraulic cleaning and fouling control in pressurized membrane process for surface water treatment. Environ. Earth Sci. **78**, 21.1–21.9 (2019)
22. S. Gao, Z. Wang, T. Ren, Y. Zhan, A combined mechanism (the open pores-cake dissolution) model for describing the trans-membrane pressure (Pb(t)) reduction in the backwash process at a constant flow rate. J. Environ. Chem. Eng. **9**, 106871 (2021)
23. G. Bolton, D. Lacasse, R. Kuriyel, Combined models of membrane fouling: development and application to microfiltration and ultrafiltration of biological fluids. J. Membr. Sci. **277**, 75–84 (2006)
24. J. Hermia, Constant pressure blocking filtration law-application to powder-law non-Newtonian fluids. **60**, 187–193 (1982)
25. A.Y. Kirschner, Y.-H. Cheng, D.R. Paul, R.W. Field, B.D. Freeman, Fouling mechanisms in constant flux crossflow ultrafiltration. J. Membr. Sci. **574**, 65–75 (2019)
26. D.J. Miller, S. Kasemset, L. Wang, D.R. Paul, B.D. Freeman, Constant flux crossflow filtration evaluation of surface-modified fouling-resistant membranes. J. Membr. Sci. **452**, 171–183 (2014)
27. Y.D. Kong, Z. Wang, Prediction of the instantaneous fouling resistance of sodium alginate during water rinsing. Chem. Eng. Res. Des. **122**, 121–131 (2017)
28. M.L. Cabero, F.A. Riera, R. Alvarez, Rinsing of ultrafiltration ceramic membranes fouled with whey proteins: effects on cleaning procedures. J. Membr. Sci. **154**, 239–250 (1999)
29. Z. Qiao, Y. Guo, Z. Wang, G. Hu, A chemically enhanced backwash model for predicting the instantaneous transmembrane pressure of flat sheet membranes in constant flow rate mode. J. Membr. Sci. **469**, 410–417 (2014)
30. Z.W. Wang, J.X. Ma, C.Y.Y. Tang et al., Membrane cleaning in membrane bioreactors: a review. J. Membr. Sci. **468**, 276–307 (2014)
31. Y.D. Kong, Z. Wang, Y. Ma et al., Theory investigation on the variation of fouling resistance during water rinsing process of the membrane fouled with sodium alginate. J. Taiwan Inst. Chem. Eng. **66**, 230–238 (2016)
32. D.W. Weyburne, New thickness and shape parameters for the boundary layer velocity profile. Exp. Therm. Fluid Sci. **54**, 22–28 (2014)
33. B.C. Sakiadis, Boundary-layer behaviour on continuous solid surfaces: II. The boundary layer on a continuous flat surface. AIChE J. **7**, 221–225 (2010)
34. P. Chen, P.A. Eggleston, Allergenic proteins are fragmented in low concentrations of sodium hypochlorite. Clin. Exp. Allergy **31**, 1086–1093 (2001)

Chemical Cleaning

Nomenclature

A_f	The surface of the pore in which protein has been detected (m^2)
A_p	The surface occupied by pores (m^2)
C	The concentration of cleaning solution (wt%)
d_e	The effective pore diameter (m)
d_o	The nominal pore diameter (m)
H	The cake height on membrane surface at certain time (m)
H_o	The cake height on membrane surface at the beginning (m)
h_k	The tortuosity factor
h_k	The Kozeny factor (1)
J	The permeate flux of the membrane ($m^3\ m^{-2}\ s^{-1}$)
J_0	The DI-water flux ($m^3 \cdot m^{-2} \cdot s^{-1}$)
J_w	The pure water flux of a virgin membrane ($m^3 \cdot m^{-2} \cdot s^{-1}$)
J_r	The flux recovery ($m^3 \cdot m^{-2} \cdot s^{-1}$)
J_{cs}	The flux of cleaning solution after cleaning ($m^3 \cdot m^{-2} \cdot h^{-1}$)
$J(t)$	The membrane flux at time t ($m^3 \cdot m^{-2} \cdot s^{-1}$)
k	A kinetic constant
k_2	The second-order rate constant—in-pore deposit swelling ($m \cdot s^{-1}$)
k_α	The second-order rate constant—cake removal ($m \cdot s^{-1}$)
k_β	The cake resistance integration constant ($m \cdot s^{-1}$)
l	Thickness of membrane (m)
m	Model parameter
N	The survival number of organisms at time *t*
N_0	The survival number of organisms at time 0
n	Model parameter
$\triangle P$	The trans-membrane pressure (Pa)
P_s	The fraction of the pore surface in which protein/polysaccharide
p_1–p_{5a}	Parameters of mathematical models of alkali cleaning
p_1–p_{6d}	Parameters of mathematical models for detergent cleaning

Z. Wang (ed.), *Modelling of the Microfiltration Process*,
https://doi.org/10.1007/978-981-99-0471-6_6

R_0	The resistance of pre-rinsed membrane (m^{-1})
R_c	The hydraulic resistance of deposit at the membrane surface (m^{-1})
$R_c(t)$	The hydraulic resistance of deposit on the membrane surface at time t (m^{-1})
R_{cp}	The resistance due to concentration polarization (m^{-1})
$R_{cp}(t)$	The resistance due to the concentration polarization (m^{-1})
R_{in}	The resistance due to in-pore fouling (m^{-1})
$R_{in}(t)$	The resistance because of in-pore fouling (m^{-1})
R_m	The hydraulic resistance of the clean membrane (m^{-1})
R_{tf}	The total resistance of fouled membrane (m^{-1})
R_{tf} (0)	The total membrane fouling resistance at the beginning of chemical cleaning operation ($t = 0$) (m^{-1})
R_{tf} (t)	The total membrane fouling resistance at time t (m^{-1})
R_{tf}	The total resistance of the fouled membrane (m^{-1})
T	Absolute temperature (K)
TMP	The trans-membrane pressure (kPa)
t	Cleaning time (s)
t	Time (min), statistical measure of parameter significance
U	P3 Ultrasil
v	Cross-flow velocity ($m \cdot s^{-1}$)
z	Negative z values show the position inside the membrane

Greek Symbols

σ	Un-swollen in-pore deposit thickness (m)
ε	Membrane porosity
η	Swelling ratio (1)
μ	Liquid viscosity (Pa·s)
μ	Permeate viscosity (Pa·s)
μ_{cs}	Dynamic viscosity of cleaning solution (Pa·s)
μ_w	Dynamic viscosity of water (Pa·s)
ω	Agitation speed of magnetic stirrer (rpm)

1 Introduction

In the actual process of membrane technology application, the micron or nanometer membrane pore size is easily blocked by foulants, and it is difficult to effectively recover the membrane filtration performance (the flux/trans-membrane pressure) only by physical cleaning technology. Therefore, it has to recover the filtration performance using chemical cleaning (Fig.1 in Chap. "Backwashing").

Chemical cleaning refers to the chemical reaction between chemical cleaning reagents and foulants on the membrane surface or in the membrane pores to overcome the binding energy between the membrane and foulants, make the foulants leave the membrane surface and membrane pores, and finally achieve the purpose of effective cleaning. Chemical cleaning is a relatively practical method to solve membrane fouling, and it usually includes six processes [1–3]: (1) Diffusion of cleaning reagent in bulk solution. (2) The cleaning reagent diffuses to the contaminated surface. (3) The cleaning agent transfers mass through the dirt layer. (4) A cleaning chemical reaction occurs. (5) The reaction products return to the two-phase interface. (6) The reaction product returns to the bulk solution. The mass transfer process from the cleaning reagent to the fouling layer may be the control step of the whole cleaning process.

The efficiency of chemical cleaning was influenced by membrane module geometry, storing conditions, cleaning agent type, cleaning agent temperature, flushing flow rate, cleaning time, trans-membrane pressure difference, cleaning agent concentration, and feed system (Fig. 1).

First, selecting appropriate chemical cleaning agents according to different foulants and membrane materials is necessary. Chemicals will not react with membranes and membrane components to damage them nor cause secondary pollution to the environment [4]. Chemical cleaning reagents mainly include acid, alkali, complexing agents, oxidants, enzymes, detergents, and surfactants.

Acid is mainly used to dissolve and remove salts and inorganic minerals. The purpose of dissolving the inorganic metal ions in the scale layer and gel layer and restoring the membrane's permeability can be achieved by reducing the pH of the solution [5, 6]. Acid cleaning using hydrochloric acid (HCl), citric acid ($C_6H_8O_7$), nitric acid (HNO_3), and sulfuric acid (H_2SO_4) is more suitable for cleaning the pore blockage fouling caused by metal salts, metal particles, and inorganic coagulants. Among them, nitric acid has been used to remove calcium precipitation in more

Fig. 1 The factors influencing on the chemical cleaning efficiency

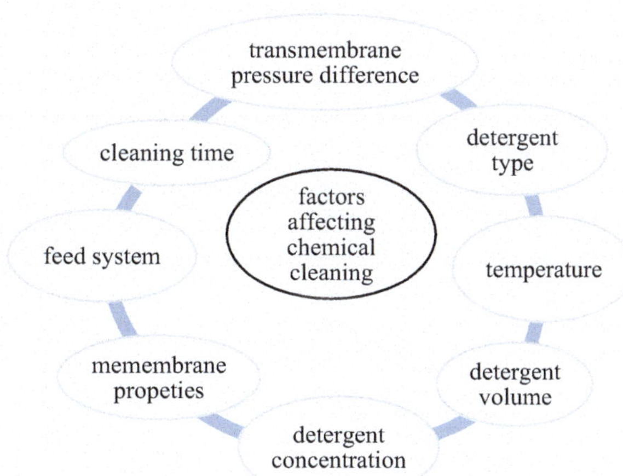

and more projects [6, 7]. In addition, nitric acid can also remove some organic and biological foulants through nitrification in some cases. However, the removal effect of acids on organic protein pollutants is poor (RRNR (relative ratio of non-removable resistance to total fouling resistance) > 20%). The cleaning effect is good when the acid concentration is high, but this will also reduce the porosity of the filter cake layer [8] and the cleaning effect (RRNR increase).

Alkaline cleaning agents, such as carbonate (sodium carbonate), phosphate (sodium tripolyphosphate), and hydroxide (sodium hydroxide), can remove protein, lipid-protein colloid, algae, and other organisms and achieve a cleaning effect by destroying the gel layer formed on the membrane surface [1, 9–11]. The effect of alkaline washing becomes better with increased cleaning time, concentration, and temperature of alkaline washing solution [12–14], and worse with increased operating pressure [12]. Sodium hydroxide, potassium hydroxide, and sodium carbonate have sound removal effects on organic protein pollutants (RRNR < 10%) [12–14]. However, the cleaning effect of saline protein wastewater is not good (RRNR > 70%) [15, 16]

Surfactants, such as sodium dodecyl sulfonate (SDS), can attract and dissolve organic substances on the membrane surface through hydrophilic and lipophilic groups [5, 17, 18]. Nonionic or anionic surfactants can be used in the cleaning process of reverse osmosis membranes; the ultrafiltration membrane for separating and treating oily sewage can be cleaned with newly prepared SDS or isoamyl alcohol solution. In addition, when using some anionic and nonionic surfactants, attention should be paid to prevent them from reacting with the membrane to cause new fouling. The scale formed by salty organic wastewater is removed effectively (RRNR = 1%) [16]. The cleaning effect becomes better with the increase in concentration. In this situation, the effect of acid or base cleaning agents or surfactants on membrane cleaning is not good. Oxidizing cleaning agents, such as hydrogen peroxide (H_2O_2), sodium hypochlorite (NaClO), and permanganate ($KMnO_4$), can be used. Oxidative cleaning agents can oxidize organics, microorganisms, protein oils, algae, and other substances, denature them and effectively separate them from the membrane surface. Among them, H_2O_2 and NaClO are the most used oxidizing cleaning agents. The former can effectively remove organics, while the latter is widely used in the membrane cleaning process. They can effectively remove organics and microorganisms and have a swelling effect on the membrane [17]. Sodium hypochlorite, hydrogen peroxide, peracetic acid, and ozone can remove organic foulants from the membrane effectively (RRNR < 10%) [14, 19]. The cleaning effect is improved with increasing cleaning time, solution concentration, and temperature. However, a long cleaning time will cause secondary pollution and reduce the cleaning effect [19]. Therefore, the material deposited in the membrane pore can be effectively cleaned.

Enzyme detergent is a selective detergent, which can effectively remove some organic substances, including protein, polysaccharide, and oil pollutants, but it will cause secondary pollution if misused. The cleaning effect is significant when the enzyme is fixed on the carrier, and the membrane is cleaned with a carrier-containing solution [13]. This method is called immobilized enzyme form, and commonly used enzymes include protease and pectinase [14].

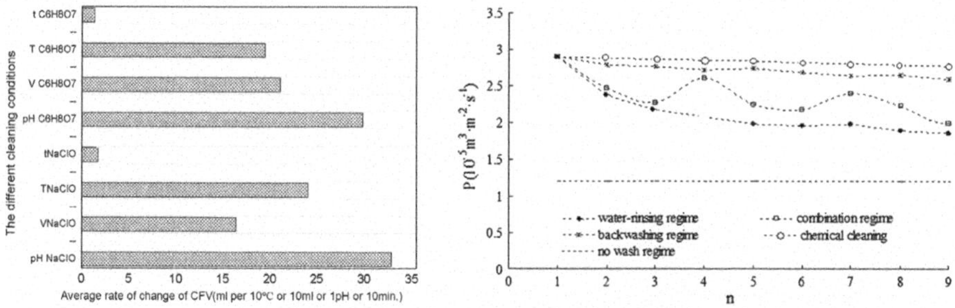

(a) The impacts of cleaning conditions

(b) The variety of the cleaning efficiency (P) for five different regimes as function of cleaning cycles (n).

Fig. 2 **a** Average rates of change of CFV for pH value of NaOCl (per 1 pH), cleaning temperature of NaOCl (per 10 °C), soaking time in NaOCl (per 10 min), volume of NaOCl (per 10 mL), pH value of $C_6H_8O_7$ (per 1 pH), cleaning temperature of $C_6H_8O_7$ (per 10 °C), soaking time in $C_6H_8O_7$ (per 10 min), and volume of $C_6H_8O_7$ (per 10 mL) and **b** the comparison of the cleaning efficiency (*P*) in different cleaning processes

The chelating agent reduces the inorganic pollutants and deposited salts adsorbed on the membrane surface and in the membrane pore by complexing with inorganic ions (such as Ca^{2+}) in the pollutants, thereby weakening membrane pollution and improving membrane flux [1, 6]. Common stinging agents include sodium ethylenediamine tetraacetate (EDTA), glucose ($C_6H_{12}O_7$), and citric acid ($C_6H_8O_7$). The cleaning effect of the membrane polluted by organic wastewater is perfect, when the cleaning time is extended from 0 to 60 min, RRNR decreased by 24% [15].

The impacts of cleaning conditions on the cleaning efficiency of 0.1 μm polyethersulfone (PES) membrane fouled by active sludge suspension from SBR were investigated systematically (Fig. 2) [20].

2 Chemical Cleaning Model

2.1 *Hou's Model [21]*

According to Darcy's law, the instantaneous membrane flux (*J(t)*, $m^3 \cdot m^{-2} \cdot s^{-1}$) can be expressed as

$$J(t) = \frac{\Delta P}{\mu R_{tf}(t)} \tag{1}$$

where ΔP is the trans-membrane pressure (Pa); μ is permeate viscosity (Pa·s); $R_{tf}(t)$ is the instantaneoustotal membrane fouling resistance (m^{-1}) and it is given by

$$R_{tf}(t) = R_m + R_{cp}(t) + R_{in}(t) + R_c\,(t) \tag{2}$$

where R_{m} is the intrinsic resistancce of the membrane (m^{-1}), $R_{\mathrm{c}}(t)$ is the hydraulic resistance of the deposit on the membrane surface (m^{-1}). $R_{\mathrm{cp}}(t)$ is the resistance due to the concentration polarization (m^{-1}) and $R_{\mathrm{in}}(t)$ is the resistance because of in-pore fouling (m^{-1}).

The weakly attached layer caused by the concentration polarization swept away during pre-rinsing [22]. Meanwhile, the particle size distribution of feed suspension is significantly larger than the mean pore size of the measured membranes (Fig. 1). Therefore, the resistance caused by the concentration polarization layer and pore blocking can be assumed to be negligible.

The relationship between the survival number of organisms and cleaning time and concentration can be expressed by the Hom-Haas model [23, 24]:

$$\ln\left(\frac{N}{N_0}\right) = -kC^m t^n \tag{3}$$

where N is the survival number of organisms at time t and N_0 is the survival number of organisms at time 0. k is a kinetic constant, C and t are cleaning agent concentration and cleaning time, respectively. m and n are model parameters.

If relating the organisms' survival number to the membrane fouling resistance, a kinetic model can be developed:

$$R_{tf}(t) = R_m + \left(R_{tf}(0) - R_m\right)e^{-kC^n t^m} \tag{4}$$

where $R_{tf}\,(t)$ is the total membrane fouling resistance at a certain chemical cleaning time t, R_{tf} (0) is the total membrane fouling resistance (m^{-1}) at the beginning of the chemical cleaning operation ($t = 0$).

Therefore, the following equation can be used to predict the membrane flux in the chemical cleaning process.

$$J(t) = \frac{\Delta p}{\mu\left(R_m + \left(R_{tf}(0) - R_m\right)e^{-kC^m t^n}\right)} \tag{5}$$

In Eq. (4), if model parameters (k, m and n) are above zero, then $R_{tf}\,(t)$ will decrease with the increase of cleaning time (t) or cleaning solution concentration (C) as shown in Fig. 3. When the cleaning time is zero, $R_{tf}\,(t)\,(t = 0)$ is equal to $R_{tf}\,(0)$. And when the cleaning time tends to infinite, $R_{tf}\,(t)\,(t \to \infty)$ will be the same value as R_m and the trend of the total membrane fouling resistance is shown in Fig. 3(a). Similarly, when the concentration of the cleaning reagent is equal to zero, $R_{tf}\,(t)\,(C = 0)$ is equal to $R_{tf}\,(0)$. When the concentration of the cleaning reagent tends to be infinite, $R_{tf}\,(t)\,(C \to \infty)$ will also be equal to R_m (Fig. 3(b)). These conclusions mentioned above are obtained as ideal conditions.

As presented in Fig. 4, the experimental results showed good agreement between the model predictions and experimental results for different fouling and cleaning

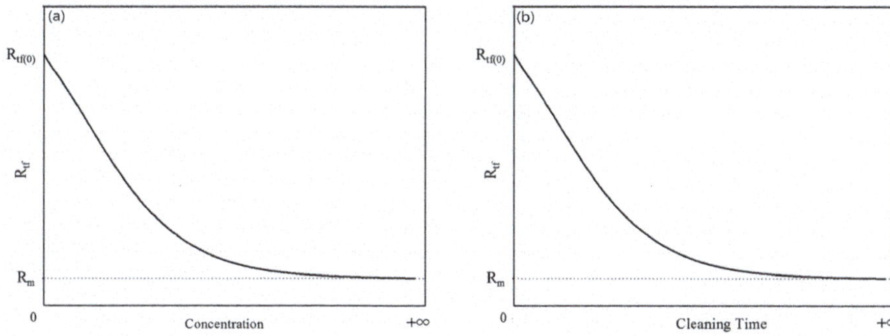

(a) $R_{tf}(t)$as function of the cleaning reagent concentration.

(b) $R_{tf}(t)$ as function of cleaning time.

Fig. 3 The theoretical analysis of the developed model

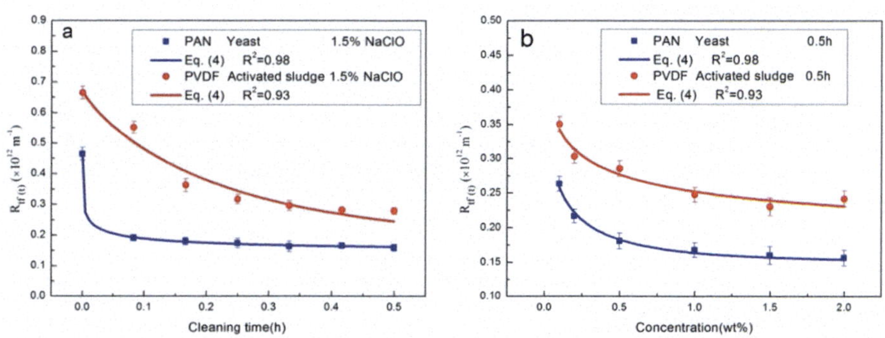

Fig. 4 The applicability of the proposed model

processes. The following results illustrated that the cleaning processes for both PAN ($R^2 = 0.98$)/PVDF ($R^2 = 0.93$) membranes fouled with yeast suspension/activated sludge suspension. The model played good agreement with NaOCl cleaning process not only for a 0.1 μm PAN membrane fouled with yeast suspension but also for a 0.1 μm PVDF membrane fouled with activated sludge suspension.

2.2 *Wang's Model [25]*

In the chemical cleaning process, the flux recovery (J_r) was evaluated by comparing the pure water flux of a fouled membrane subjected to the cleaning procedures (J_{cw}) with the pure water flux of a virgin membrane (J_{w}) as follows:

$$J_r = \frac{J_{CW}}{J_W} \times 100\% \tag{6}$$

If relating the organisms' survival number to the membrane fouling resistance, a kinetic model can be developed:

$$\ln\left(\frac{R_{tf}(t) - R_m}{R_{tf}(0) - R_m}\right) = -kC^m t^n \tag{7}$$

where R_{tf} (t) is the total fouling resistance of microfiltration membrane at a certain chemical cleaning time t, C and t are cleaning solution concentration and cleaning time, respectively.

After combining Eqs. (6) and (7):

$$J_r = \frac{1}{1 + \frac{J_w - J_o}{J_0}\exp(-kC^m t^n)} \tag{8}$$

where J_0 is the DI-water flux of the fouled membrane after the filtration operation, and m and n are empirical constants.

This model was proposed based on the Hom-Haas model and it can accurately describe the change of flux recovery rate (J_r) with cleaning time (t) and cleaning agent concentration (C) in the chemical cleaning process for sodium hypochlorite, sodium hydroxide, and sodium dodecyl sulfonate (Fig. 5). The model parameters m and n are only related to the operating pressure and temperature, while the operating pressure determines the rate constant k, temperature, and stirring speed.

2.3 *Bird's Model [26]*

The instantaneous membrane flux ($J(t)$, $m^3 \cdot m^{-2} \cdot s^{-1}$) in the chemical cleaning process can be expressed as Eq. (1), and the total membrane fouling resistance ($R_{tf}(t)$) (m^{-1}) can be expressed as follows due to the concentration polarization phenomenon was found to play a minor role:

$$R_{tf}(t) = R_m + R_{in}(t) + R_c(t) \tag{9}$$

where R_m is the resistance of membrane itself (m^{-1}), R_{in} is the resistance due to in-pore fouling (m^{-1}), and R_c is the cake resistance of the deposit at the membrane surface (m^{-1}).

Suppose that R_c has been modeled as a simple second-order resistance decrease as the cake is cleaned:

$$\frac{dR_c}{dt} = -k_\alpha R_c^2 \tag{10}$$

After integration:

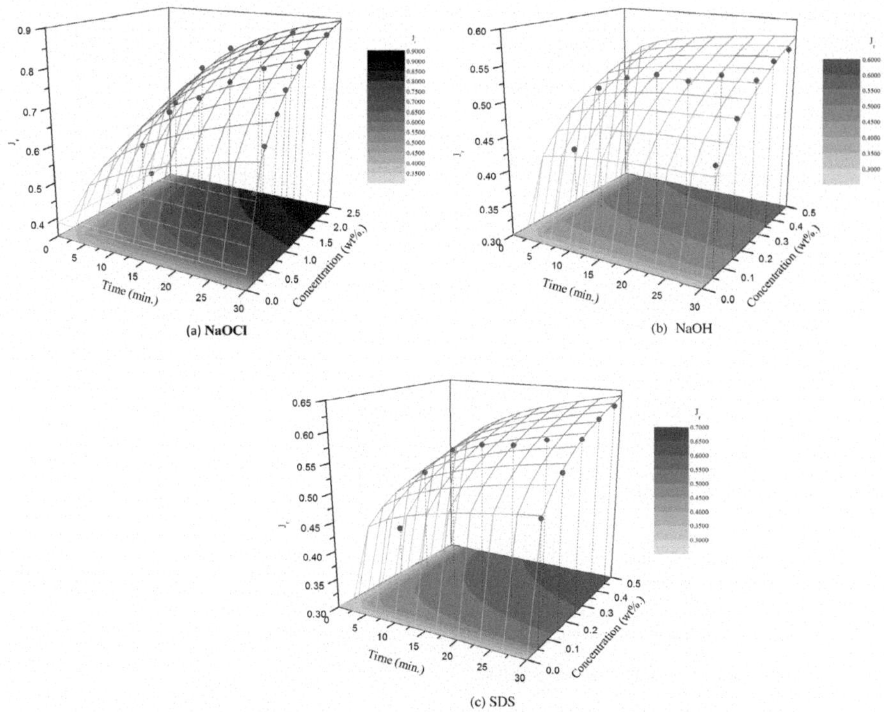

Fig. 5 Comparison between experimental data and model predictions in terms of flux recovery versus cleaning time and concentration for cleaning agent of **a** NaOCl, **b** NaOH and **c** SDS (0 MPa, 400 rpm, and 28 °C) [25]

$$R_c = 1/\left(k_\alpha t + \frac{1}{R_c(0)}\right) \tag{11}$$

R_i is the resistance due to in-pore fouling (m^{-1}) and it can be calculated by a modified Carmen-Kozeny equation:

$$R_i = \frac{36h_k(1-\varepsilon)^2 l}{\varepsilon^3 d_e^2} \tag{12}$$

where h_k is the tortuosity factor, ε is the porosity, l is the membrane thickness (m) and d_e is the effective pore diameter (m), which can be given by the nominal pore diameter d_o minus twice the swollen in-pore deposit thickness:

$$d_e = \left(d_0 - 2\left[\left(\frac{\delta}{k_2 t + \delta}\right)(\eta k_2 t + \delta)\right]\right) \tag{13}$$

Combining (11), (12), and (13) into (9):

Fig. 6 The comparison of model predictions and experimental data for the resistance changing with time for the cleaning of a 2.0 μm sintered stainless steel membrane cleaned using 0.2 wt% sodium hydroxide at 50 °C, TMP 0.5 bar, CFV 1.6 m/s [26]

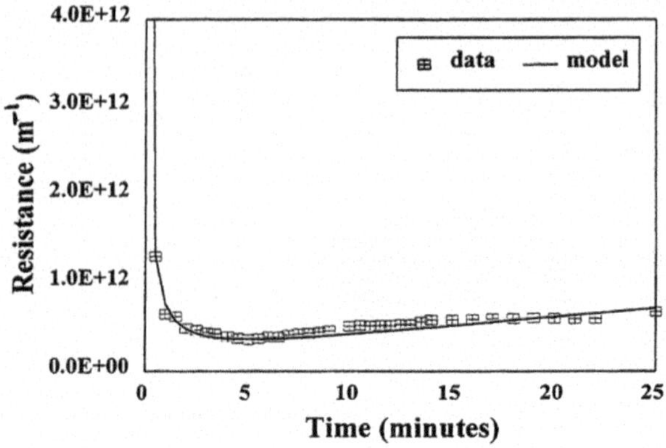

$$R_t(t) = 1/\left(k_\alpha t + \frac{1}{R_c(0)}\right) + \frac{36h_k(1-\varepsilon)^2 l}{\varepsilon^3\left(d_0 - 2\left[\left(\frac{\delta}{k_2 t+\delta}\right)(\eta k_2 t + \delta)\right]\right)^2} + R_m \tag{14}$$

It can describe the unsteady-state hydraulic resistance variation that occurs when surface and in-pore bound material undergoes morphological changes during caustic cleaning. The model relies on simultaneous removal and swelling processes occurring within the surface and in-pore deposits.

Therefore, the instantaneous membrane flux ($J(t)$) in the chemical cleaning process can be expressed as (Fig. 6)

$$J(t) = \frac{\Delta p}{\mu\left(1/\left(k_\alpha t + \frac{1}{R_c(0)}\right) + \frac{36h_k(1-\varepsilon)^2 l}{\varepsilon^3\left(d_0-2\left[\left(\frac{\delta}{k_2 t+\delta}\right)(\eta k_2 t+\delta)\right]\right)^2} + R_m\right)} \tag{15}$$

2.4 *Popović's Model [22]*

A ceramic tube membrane separated the whey protein solution, and the fouled membrane was cleaned: the cleaning scheme is to conduct a 30 min hydraulic flushing, followed by a 30 min chemical cleaning, and finally, a 30 min hydraulic flushing. The two selected chemical cleaning agents are alkali (NaOH) and detergent (P3 ultrasil).

Since the fouled membrane is rinsed with water before being exposed to chemical cleaning, it can be assumed that a weakly attached layer due to concentration polarization will be swept away during pre-rinsing.

Suppose that the cake resistance (R_c) has been modeled as a simple first-order resistance decrease as the cake is cleaned:

$$\frac{dR_c}{dt} = -p_1 R_c \rightarrow R_c = \exp(-p_1 t + p_2) \quad (16)$$

The swelling kinetics for in-pore bound material were expressed alternatively:

(a) as a function of the second order, the following expression can be obtained for alkali cleaning,

$$R_{in} = p_{3a}t^2 + p_{4a}t + p_{5a} \quad (17)$$

(b) as a function of the third order, the following expression can be obtained for detergent cleaning:

$$R_{in} = p_{3d}t^2 + p_{4d}t + p_{5d} \quad (18)$$

After substituting Eqs. (16), (17), (18) into (9):

For alkali cleaning:

$$R_{tf} = \frac{TMP}{J_w \mu_w} + \exp(-p_1 t + p_2) + p_{3a}t^2 + p_{4a}t + p_{5a} \quad (19)$$

For detergent cleaning:

$$R_{tf} = \frac{TMP}{J_w \mu_w} + \exp(-p_1 t + p_2) + p_{3d}t^2 + p_{4d}t + p_{5d} \quad (20)$$

Effective pore diameter and thickness of in-pore bound material.

The relationship between the R_{in}-resistance and the effective pore diameter (d_e) can be established using the modified Carmen-Kozeny equation:

$$R_i = \frac{36h_k(1-\varepsilon)^2 l}{\varepsilon^3 d_e^2} = kd_e^{-2} \quad (21)$$

where the parameter k covers a tortuosity factor (h_k), membrane porosity (ε), and membrane thickness equal to the pore length (l).

From Eqs. (17), (18), and (21):

For alkali cleaning:

$$p_{3a}t^2 + p_{4a}t + p_{5a} = k_a d_e^{-2} \quad (22)$$

For detergent cleaning:

$$\left(p_{3d}t^2 + p_{4d}t + p_{5d}\right) = k_d d_e^{-2} \quad (23)$$

The parameters k_a and k_d can be evaluated from Eqs. (22) and (23), using the conditions achieved at the end of the cleaning process.

For alkali cleaning:

$$(p_{3a}t^2 + p_{4a}t + p_{5a})d_e^2 = k_a \tag{24}$$

For detergent cleaning:

$$\left(p_{3d}t^2 + p_{4d}t + p_{5d}\right)d_e^2 = k_d \tag{25}$$

The effective diameter of pores at the end of cleaning ($d_e(t_{end})$) is related to the cleaning efficiency, which can be expressed as a ratio of the flux of cleaning solution through membrane at the end of chemical cleaning (J_{Cs}) to the water flux through a new membrane (J_W):

$$J_r = \frac{J_{Cs}}{J_W} \times 100\% \tag{26}$$

In fact, the flux recovery (J_r) can be also expressed as the ratio of the filtration surface available at the end of cleaning ($S_e(t_{end})$ to the nominal filtration surface (S_0), as follows:

$$\frac{S_e(t_{end})}{S_0} = \frac{d_e^2(t_{end})}{d_0^2} = J_r \rightarrow d_e(t_{end}) = J_r^{0.5}d_0 \tag{27}$$

Obviously, it is assumed that the foulants accumulated into the volume of pores after cleaning are attached to the surface of pores, thus reducing their nominal diameter (d_o) to the effective diameter at the end of cleaning ($d_e(t_{\mathrm{end}})$). The effective diameter can thus be expressed as a function of both the cleaning efficiency and the nominal diameter. After estimation of $d_\mathrm{e}(t_\mathrm{end})$ using Eq. (25), the parameters k_a and k_d can be calculated from Eqs. (24) and (25).

Knowing k-values and using Eqs. (22) and (23), the following equations can be obtained:

For alkali cleaning:

$$d_e(t) = \sqrt{\frac{k_a}{p_{3a}t^2 + p_{4a}t + p_{5a}}} \tag{28}$$

For detergent cleaning:

$$d_e(t) = \sqrt{\frac{k_d}{p_{3d}t^2 + p_{4d}t + p_{5d}}} \tag{29}$$

The thickness of the layer attached to the wall of pores ($\delta(\mathrm{t})$) at any time in the cleaning process:

$$d_e(t) = d_0 - 2\delta(\mathrm{t}) \rightarrow \delta(\mathrm{t}) = \frac{d_0 - d_e(t)}{2} \tag{30}$$

The proposed models can be used to predict the variation of the total and specific resistances. Moreover, the models also can be used to estimate the pore diameter and related thickness of the in-pore deposit change with time in the alkali and detergent cleaning processes (Fig. 7).

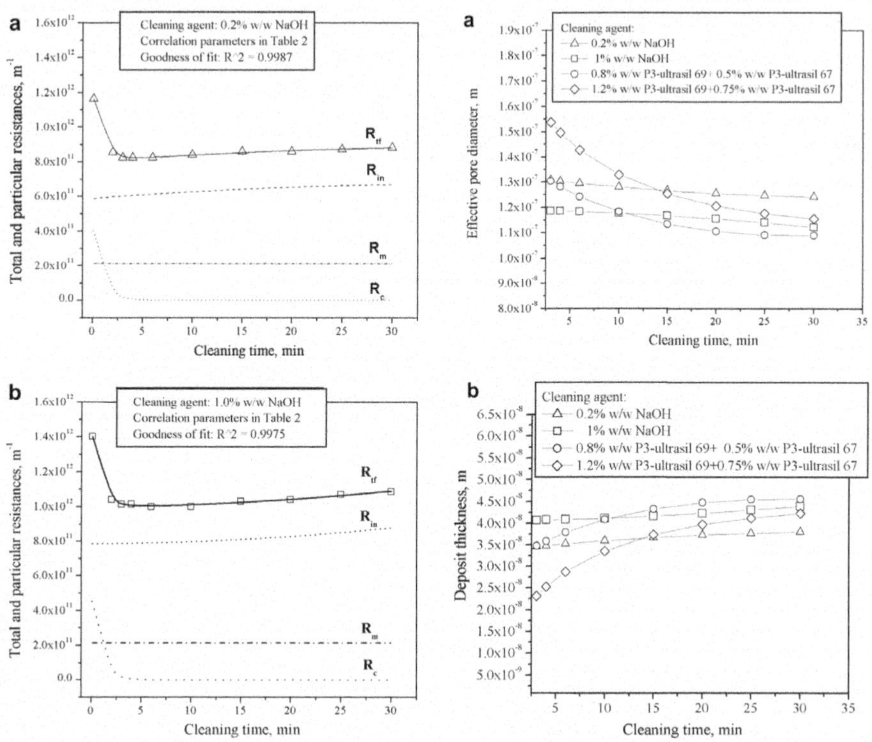

(a) Resistances during cleaning with 0.2%w/w NaOH and 1.0%w/w NaOH, at 50ºC, TMP 30 kPa and CFV 1.73 m/s; (R_{tf}–total resistance, R_m – membrane resistance, R_c – cake resistance, R_{in}–in-pore fouling resistance).

(b) Effective pore diameter and thickness of the in-pore deposit during alkali and detergent cleaning

Fig. 7 The variation of the resistances (R_{tf}, R_{m}, R_{c}, and R_{in}), the effective pore diameter (d_e), and the thickness of the in-pore deposit in the alkali and detergent cleaning process ($\delta(t)$) [22]

2.5 Zator's Model [27]

A confocal scanning laser microscope (CSLM) and visible results (the permeate flux evolution) were used to evaluate the effectiveness of water rinsing and chemical cleaning protocols on 0.8 μm polycarbonate membranes fouled with model solutions containing BSA and dextrans. Water rinsing for 30 min and chemical cleaning cycles of 5, 15, and 30 min with two different concentrations of an enzymatic cleaning agent (P3 Ultrasil 53) were tested. The results showed that water rinsing does not improve membrane permeability if the protein is involved in membrane fouling. Some of the proteins are driven deeper inside the membrane, causing further blockage and reducing water flux in the water rinsing process, while as for chemical cleaning protocols, the cleaning operating conditions (concentration of the cleaning agent and cleaning time) resulted in the highest water permeate flux recovery were determined. P_s profiles and CSLM images showed that most of the protein and the dextran were removed from the membrane during chemical cleaning.

The surface occupied by pores (A_p) and the fraction of the pore surface in which protein/polysaccharide was detected (P_s), which can be expressed as [28]

$$P_s = \frac{A_f}{Ap} \times 100\% \tag{31}$$

where A_f is the surface of the pore in which protein has been detected.

In order to obtain more quantitative data from CSLM analysis, average P_s values were fitted versus z values using an empirical equation (Eq. 27) assuming protein adsorption inside the pores.

$$P_s = \frac{P_{s0}}{1 + e^{-[(z - z_{1/2})/k]}} \times 100\% \tag{32}$$

where P_{s0}: P_s value at the membrane surface and $z_{1/2}$: z value (depth from the surface) where $P_s = P_{s0}/2$ (Fig. 8).

2.6 Koyuncu's Model [29]

Based on the Hom-Haas model [23, 24] and considering the variation of the cake removal as a function of the chemical cleaning agent concentration and cleaning time in the chemical cleaning process, the following equation can be obtained:

$$\ln\left(\frac{H}{H_0}\right) = -kC^n t^m \tag{33}$$

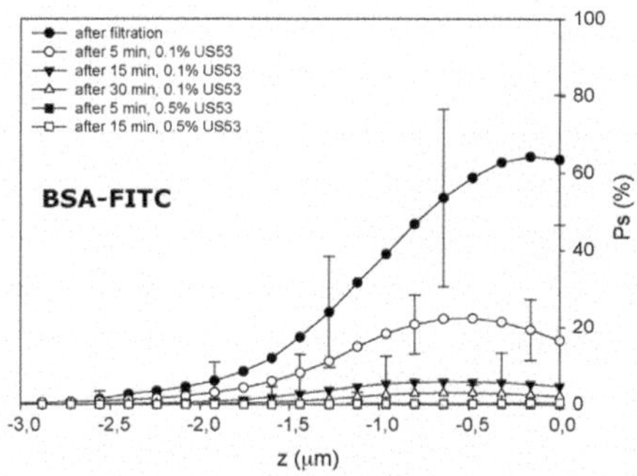

(a) Fraction of pore surface on which BSA–FITC is detected (P_s) after filtration and cleaning steps of different lengths using binary BSA–FITC/dextran–RITC 70kDa 0.5g/L solutions with two concentrations of US53; negative *z* values show the position inside the membrane. Solid curve: fitting to Eq. (5-27)

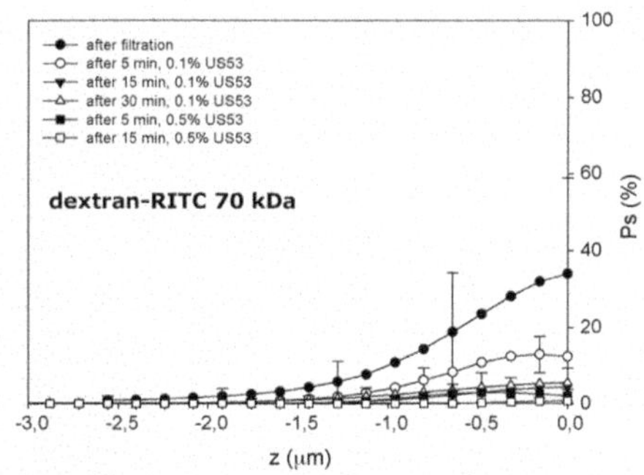

(b) Fraction of pore surface on which dextran–RITC is detected (P_s) after filtration and cleaning steps of different lengths using binary BSA–FITC/dextran–RITC 70kDa 0.5g/L solutions with two concentrations of US53; negative *z* values show the position inside the membrane. Solid curve: fitting to Eq. (5-27).

Fig. 8 Dextran–RITC P_s values obtained after filtration and after five different cleaning cycles [27]

where H is cake height on membrane surface at a certain time, H_o is cake height on the membrane surface at the beginning, *k, n* and *m k* is model parameters, C is the concentration of cleaning solution and t is the cleaning time.

From Eq. (28):

$$H = \frac{H_0}{e^{kC^n t^m}} \tag{34}$$

The cake layer height between the cleaned and uncleaned side of the membranes fouled with $CaCl_2$ and humic acid solution before and after the chemical cleaning operation was measured by a vertical scanning interferometer (VSI). The results showed that the model predictive cake heights were in good agreement with measured cake heights by Vertical scanning interferometer (VSI) for (a) H_2O_2 and (b) NaCl solutions (Figs. 9 and 10).

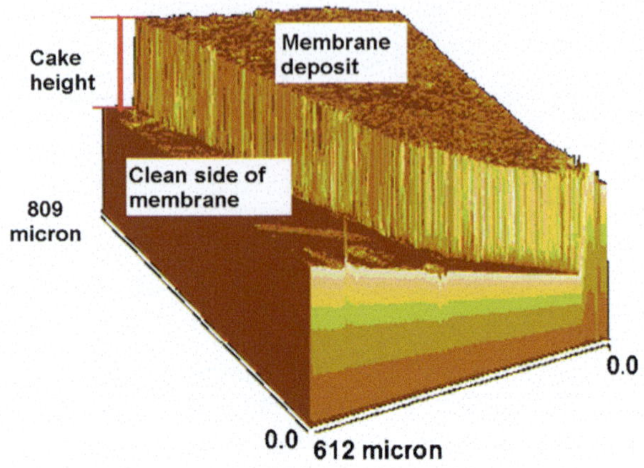

Fig. 9 **a** Interferometric measurement of three-dimensional height differences between cleaned and uncleaned side of membrane sample [29]

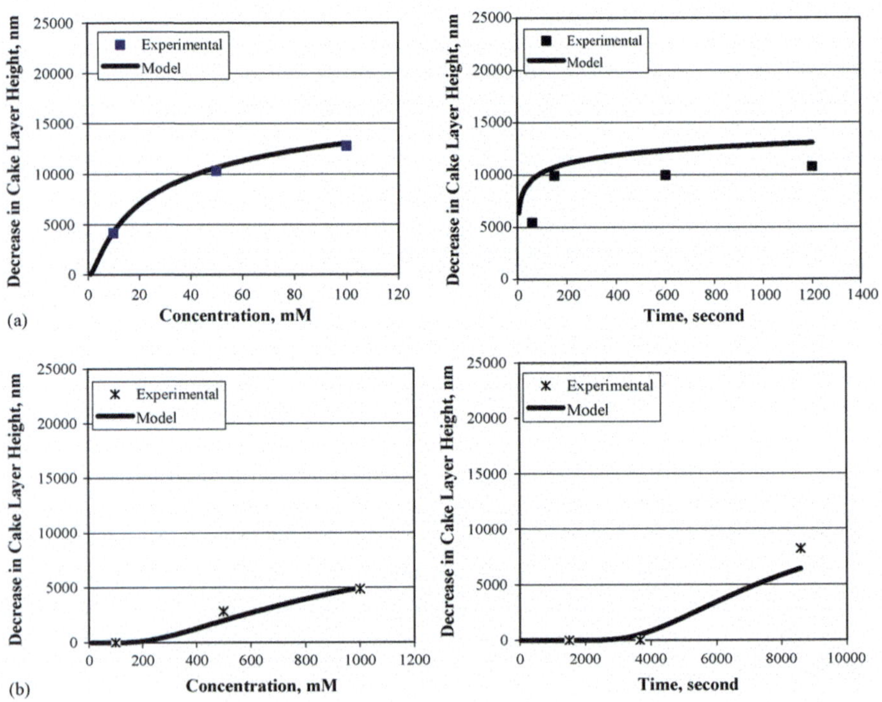

Fig. 10 Comparison of the model predictions and measured values for **a** H_2O_2 and **b** NaCl solutions [29]

References

1. N. Porcelli, S. Judd, Chemical cleaning of potable water membranes: a review. Sep. Purif. Technol. **71**, 137–143 (2010)
2. C. Shorrock, M. Bird, Membrane cleaning: chemically enhanced removal of deposits formed during yeast cell harvesting. Food Bioprod. Process. **76**, 30–38 (1998)
3. E. Plett, Cleaning of fouled surfaces, in *Foulingand Cleaning in Food Processing*. ed. by D. Lund, E. Plett, C. Sandu (University of Wisconsin-Madison Ext Dup, Wisconsin, 1985), p.286
4. S.S. Madaeni, S. Samieirad, Chemical cleaning of reverse osmosis membrane fouled by wastewater. Desalination **257**, 80–86 (2010)
5. G. Tra¨gardh, Membrane cleaning. Desalination **71**, 325–335 (1989)
6. L.J. Zeman, A.L. Zydney, *Microfiltration and Ultrafiltration: Principles and Applications* (Dekker, New York, 1996)
7. N.M. D'Souza, A.J. Mawson, Membrane cleaning in the dairy industry: a review. Crit. Rev. Food Sci. Nutr. **45**, 125–134 (2005)
8. S.S. Madaeni, Y. Mansourpanah, Chemical cleaning of reverse osmosis membranes fouled bywhey. Desalination **161**, 13–24 (2004)
9. H. Lee, G. Amy, J. Cho, Y. Yoon, S.-H. Moon, I.S. Kim, Cleaning strategies for fluxrecovery of an ultrafiltration membrane fouled by natural organic matter. Water Res. **35**, 3301–3308 (2001)
10. A. Al-Amoudi, R.W. Lovitt, Fouling strategies and the cleaning system of NF membranes and factors affecting cleaning efficiency. J. Membr. Sci. **303**, 4–28 (2007)
11. M. Cheryan, Ultrafiltration Handbook. Technomic Publishing Company, Inc. (1986)
12. M. Bartlett, M.R. Bird, J.A. Howell, An experimental study for the development of a qualitative membrane cleaning model. J. Membr. Sci. **105**, 147–157 (1995)
13. S.S. Popović, M.N. Tekić, M.S. Djurić, Kinetic models for alkali and detergent cleaning of ceramic tubular membrane fouled with whey proteins. J. Food Eng. 94, 307–315 (2009)
14. M.F. Rabuni, N.M. NikSulaiman, M.K. Aroua, C. YernChee, N. AwanisHashim, Impact of in situ physical and chemical cleaning on PVDF membrane properties and performances. Chem. Eng. Sci. **122,** 426–435 (2015)
15. W.S. Ang, S. Lee, M. Elimelech, Chemical and physical aspects of cleaning of organic-fouled reverse osmosis membranes. J. Membr. Sci. **272**, 198–210 (2006)
16. Q. Li, M. Elimelech, Organic fouling and chemical cleaning of nanofiltration membranes: measurements and mechanisms. Environ. Sci. Technol. **38**, 4683–4693 (2004)
17. M.F. Cox, Detergents and Cleaners: A Handbook for Formulators (1994)
18. N.G. Marriott, R.B. Gravani, *Principles of Food Sanitation* (Springer, New York, 2006)
19. V. Puspitasari, A. Granville, P. Le-Clech, V. Chen, Cleaning and ageing effect of sodium hypochlorite on polyvinylidene fluoride (PVDF) membrane. Separ. Purif. Technol. **72,** 301–308 (2010)
20. Z. Wang, Q. Zhang, Y. Ma, L. Yang, Y. Lyu, S. Zhao, Influence of various operating conditions on cleaning efficiency in sequencing batch reactor (SBR) activated sludge process. Part III: Chem. Clean. Desalin. **325**, 122–131 (2013)
21. L. Hou, K. Gao, P. Li, X. Zhang, Z. Wang, P. Song, W. Yao, A kinetic model for calculating total membrane fouling resistance in chemical cleaning process. Chem. Eng. Res. Des. **128**, 59–72 (2017)
22. S.S. Popovicí, M.N. Tekic, M.S. Djuric, Kinetic models for alkali and detergent cleaning of ceramic tubular membrane fouled with whey proteins. J. Food Eng. **94**, 307–315 (2009)
23. L.W. Hom, Chlorine disinfection of ponded ecosystems. Chimie Ind. Gen. Chimique **105**, 33 (1972)
24. C.N. Haas, J. Joffe, Disinfection under dynamic conditions-modification of Hom model for decay. Environ. Sci. Technol. **28**, 1367–1369 (1994)
25. Z. Wang, X. Zhang, Z. Zhu, Y. Kong, K. Gao, W. Yao, Influence of various operating conditions on cleaning efficiency in sequencing batch reactor (SBR) activated sludge process. Part V: Chemical cleaning mode. J. Taiwan Inst. Chem. Eng. **63**, 52–60 (2016)

26. M.R. Bird, M. Bartlett, Measuring and modelling flux recovery during the chemical cleaning of MF membranes for the processing of whey protein concentrate. J. Food Eng. **53**, 143–152 (2002)
27. M. Zator, J. Warczok, M. Ferrando, F. López, C. Güell, Chemical cleaning of polycarbonate membranes fouled by BSA/dextran mixtures References: J. Membr. Sci. **327**, 59–68 (2009)
28. M. Ferrando, A. Rˇozek, M. Zator, F. López, C. Güell, An approach to membrane fouling characterization by confocal microscopy. J. Membr. Sci. **250**, 283–293 (2005)
29. I. Koyuncu, A. Luttge, M.R. Wiesner, Interferometric observations and kinetic modeling of the chemical cleaning of humic materials deposited on membranes. J. Membr. Sci. **313**, 127–134 (2008)

Ultrasonic Cleaning

Nomenclature

$A\ (t)$	The instantaneous available frontal membrane area in the membrane filtration process (m^2)
A_f	The available frontal membrane area after the membrane filtration process (m^2)
$FR(t)$	The instantaneous flux recovery
b_0	The constant coefficient
b_{i}	The linear coefficient
b_{ii}	The quadratic coefficient
b_{ij}	The interaction coefficient
C	The concentration of the cleaning agent (kg/m^3)
CE	The ultrasonic cleaning efficiency
CFV	The cross-flow velocity (m/s)
J_0	Pure water flux of the virgin membrane or rinsed membrane (m/s)
$J(t)$	The instantaneous membrane flux (m/s)
J_{fc}	The membrane permeate flux after ultrasonic cleaning (m/s)
R_{tf}	The membrane fouling resistance (m/s)
K	The steady membrane frontal area (m^2);
K_b	Complete blocking resistance constant (h^{-1})
K_C	Cake resistance constant (hm^{-2});
k	Kinetic constant
n	The number of factor or experimental constant
PL	Ultrasound power level
$\triangle P$	The trans-membrane pressure (Pa)
$R_f\ (0)$	Total fouling resistance at rinsing time of 0 s (m^{-1})
$R_f\ (t)$	Total fouling resistance at rinsing time of t (m^{-1})
R_m	The virgin membrane resistance (m^{-1})
R_{fC}	The residual cake resistance after the ultrasonic cleaning (m^{-1}).
R_{pore}	The pore resistance (m^{-1})

Z. Wang (ed.), *Modelling of the Microfiltration Process*,
https://doi.org/10.1007/978-981-99-0471-6_7

$R_{pB\,(t)}$	The membrane pore resistance due to pore blocking (m^{-1})
$R_{\_pin\,(t)}$	The membrane pore resistance due to the foulants depositing and narrowing the membrane pore (m^{-1})
Rtf	Total fouling resistance (m^{-1})
S	Membrane surface (m^2)
TMP	Trans-membrane pressure (Pa)
UFM	Ultrasound frequency mode
t	Ultrasonic cleaning time (s)
x_i	The input factor
x_j	The input factor
y	The response variable

Greek Symbols

μ	Liquid viscosity (Pa s)

1 Introduction

Ultrasonic cleaning method has the characteristics of high speed and environmental friendliness and attracts more and more people's attention. Ultrasonic (US) cleaning could be an alternative method for membrane cleaning compared to traditional physical and chemical methods (Fig. 1). The cavitation effect of the ultrasonic wave can enhance the washing effect and improve cleaning efficiency. In addition, compact, less land occupation and other make the application of the ultrasonic wave more and more widely and ultrasonic cleaning is also used in conjunction with other cleaning processes, such as chemical cleaning, to improve cleaning effectiveness (Fig. 2).

The ultrasonic cleaning process uses cavitation bubbles, microbeams, microflows, microjets and shock waves generated by ultrasound to remove pollutants on the membrane. However, it does not affect foulants in the membrane pores [1–6]. The ultrasonic cleaning efficiency is influenced by ultrasonic frequency, intensity, external pressure, liquid temperature, the liquid property, gas, duty cycle, ultrasonic duration, and ultrasonic power (Fig. 3). Few bubbles are generated by low-frequency ultrasound. However, cavitation bubbles can obtain greater amplitude [7, 8], so they can remove pollutants on the membrane surface more effectively than high-frequency ultrasound [1, 9]. Mikko et al. found that the relative ratio of non-removable resistance to total fouling resistance (RRNR) decreased by 22% when the frequency increased from 70 to 620 kHz [8]. In addition, a better cleaning effect can be obtained by combining the use of high frequency to generate more cavitation bubbles with the use of low frequency to make bubbles obtain greater amplitude [10–12]. Maskooki et al. found that mixed frequency (28, 45, and 100 kHz) cleaning (RRNR = 14%)

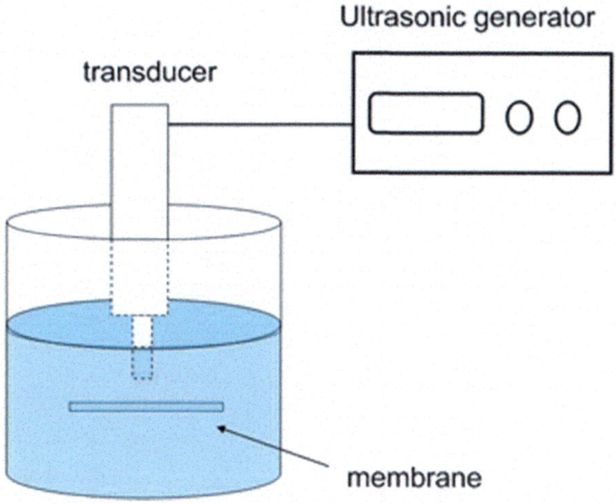

Fig. 1 The Schematic diagrams of the ultrasonic generator

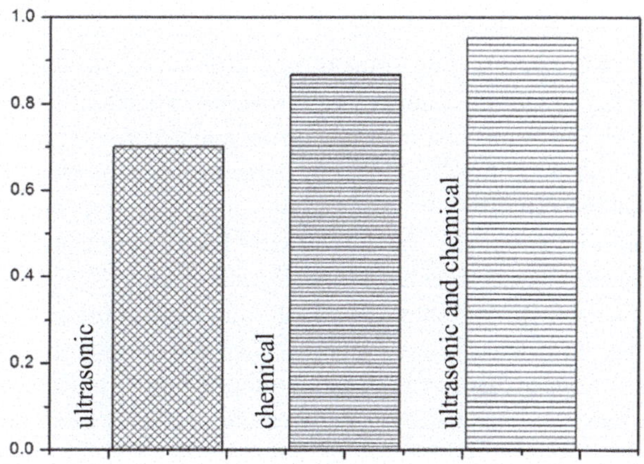

Fig. 2 Comparison of the flux recovery between ultrasonic cleaning, chemical cleaning, and combined ultrasonic and chemical cleaning method (soaking in 0.3 wt% NaClO and 15 wt% citric acid 1 h, respectively; 25 °C)

is better than single frequency (18% (28 kHz), 24% (45 kHz), 36% (100 kHz)), the cleaning effect increased by 4%, 10%, and 22%, respectively. The higher the energy density, the more cavitation bubbles, the larger the cavitation area, and the better the corresponding cleaning effect [10]. It is worth noting that excessive strength may damage the membrane structure. When Kan et al. [13] raised the ultrasonic power from 0 to 15W, the RRNR decreased from 12.8% to 1.8%. The cleaning effect can be improved by increasing the ultrasonic treatment time, but the membrane will be damaged if the treatment time is too long. Lamminen et al. [14] found that ultrasonic cleaning of PVDF membranes polluted by sulfuric acid latex particles. A good cleaning effect can be achieved in two minutes, and it will remain unchanged after that. The cleaning effect is also related to the intermittent pause time. Kan et al. [13] found that when ultrasonic cleaning of bentonite and humic acid contaminated PTFE membranes, RRNR for the intermittent time of 9 s and 0 s were 13.7% and

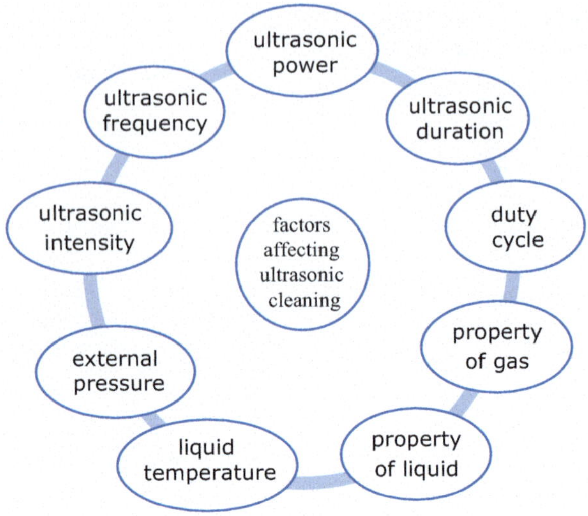

Fig. 3 The factors influencing on the ultrasonic cleaning efficiency

12.8%, respectively. Temperature affects the saturated vapor pressure and viscosity of liquid and the intensity of ultrasonic wave and cavitation threshold. Li et al. [15] achieved better results when ultrasonic low-temperature cleaning of nylon microfiltration membranes polluted by paper mill wastewater. However, Chai et al. [16] found that the effect of ultrasonic high-temperature cleaning of PS membrane polluted by peptone was good. The properties of feed solution and membrane materials also affect the ultrasonic cleaning effect: Luján et al. [17] found that the ultrasonic cleaning effect on BSA-contaminated ceramic membrane is better than PES flat membrane. The ultrasonic cleaning effect is poor at the higher feed concentration. The increased operating pressure will cause cavitation bubble collapse, reduction, and poor cleaning effect. Shobha et al. [18] reduced the operating pressure from 200 to 50 kPa, and the corresponding RRNR decreased from 57 to 32%.

It should be stated that the ultrasonic devices on the lab scale usually have two categories (Fig. 4). One refers that the ultrasonic probe directly irradiates the foulant layer, as shown in Fig. 4a. In this situation, the distance between the ultrasonic probe and the foulant layer also becomes an essential factor in influencing ultrasonic cleaning efficiency. The distance range of 2–8 cm satisfied operation conditions for most devices [13, 19–21]. Another device is the ultrasonic transducer attached to the ultrasonic cleaning bath, shown in Fig. 4b. However, if the feed is filled with the bath, we call it a direct ultrasonic cleaning device. If other liquid media is filled in the bath and the membrane device is soaked into this liquid, we may call it an indirect ultrasonic cleaning device.

The ultrasonic cleaning mechanism, including acoustic streaming (Fig. 5a), microstreaming (Fig. 5b), microstreamers (Fig. 5c), and microjet (Fig. 5d) [8]. Acoustic streaming allows liquid flow from centimeters to tens of centimeters, which can remove the loose foulant layers. Microstreaming means cavitation bubbles oscillating in the ultrasonic field, which brings about their surrounding liquid

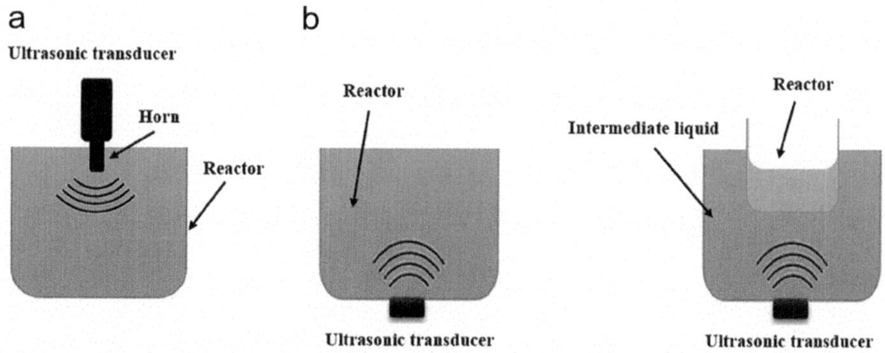

Fig. 4 **a** Direct ultrasonic irradiation with an ultrasonic probe. **b** Direct (left) and indirect (right) ultrasonic irradiation with ultrasonic bath

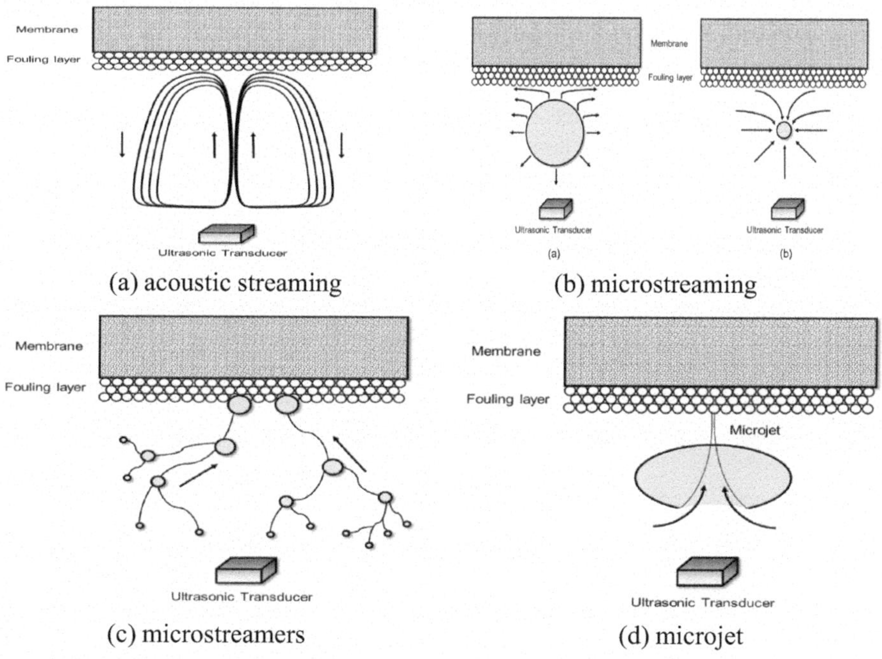

Fig. 5 The ultrasonic cleaning mechanism

flow. Usually, this effect range is about a micron range. The following term, microstreamers, is different from microstreaming. Microstreamers refer to bubbles that travel in ribbon-like structures, which can cause an adequate range of millimeters. Micro-jet with ultrahigh speed (100 ~ 200 m/s) and pressure (above 100 MPa) is created when a cavitation bubble collapses. However, the effective range of microjet is only on the micron order.

2 Ultrasonic Cleaning Model

2.1 Alventosa-Delara's Model [11]

The factors (trans-membrane pressure (*TMP*), cross-flow velocity (*CFV*), ultrasound power level (*PL*), and ultrasound frequency mode (*UFM*)) affecting the ultrasound-assisted cleaning of an ultrafiltration ceramic membrane fouled by dye particles was carried out.

A Box-Behnken Design was used to find the optimal conditions for the cleaning procedure through a response surface study. The following statistical mode was used to determine the mathematical relationship between the independent variable and the dependent variable:

$$y = b_0 + \sum_{i=l}^{n} b_i x_i + \sum_{i=l}^{n} b_{ii} x_i^2 + \sum_{i=l}^{n-l} \sum_{j=i+l}^{n} b_{ij} x_i x_j \tag{1}$$

where y is the response variable, x_{i} and x_{j} are the input factors, n is the number of factor, b_0 is the constant coefficient, b_{i} is the linear coefficient, b_{ii} is the quadratic coefficient, and b_{ij} is the interaction coefficient. Multiple regression analysis is performed to obtain the coefficients and the equation can be used to predict the response variable.

The result of the statistical data analysis is presented in a Standardized Pareto chart (Fig. 6), which is a graphical representation of each of the effects (linear, quadratic, and interactions), plotted in decreasing order of magnitude. The length of the bars is proportional to the absolute magnitude of the estimated coefficients. The equation describing the response (Cleaning efficiency (y)) is obtained:

$$\begin{aligned} y =& 13.3790 + 7.3355\, TMP + 3.5639\, CFV \\ &+ 0.0413\, Power\ level - 3.3680\, TMP^2 \end{aligned} \tag{2}$$

where the values of the variables are specified in their original units (Fig. 7).

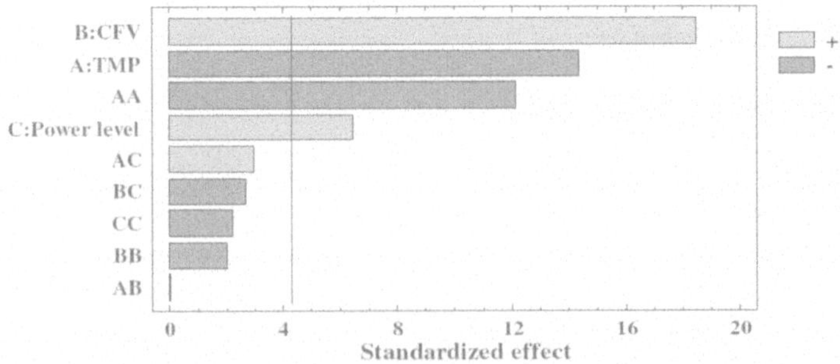

Fig. 6 Standardized Pareto chart for the analysis of cleaning efficiency when using mixed wave sonication mode

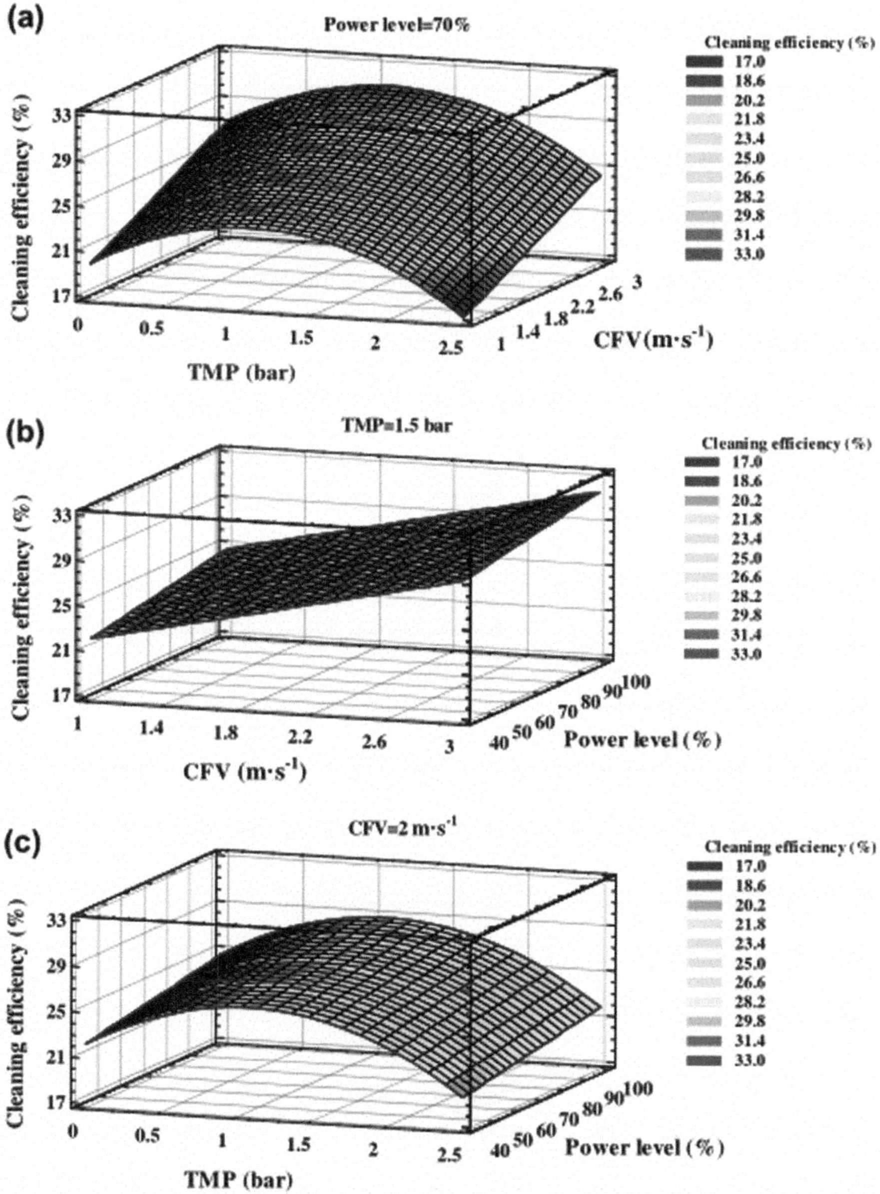

Fig. 7 3D response surface plot shows the effect of **a** TMP and CFV **b** CFV and power level, **c** TMP and power level on cleaning efficiency when using mixed wave sonication mode (Statistical model for the predicted data: cleaning efficiency)

2.2 Luo's Model [22]

According to Darcy la [22]:

$$FR(t) = \frac{J_{fc}}{J_0} \times 100\% = \frac{R_m + R_{tf}(t)}{R_m + R_{tf}(0)} \times 100\% \tag{3}$$

where FR(t) is the instantaneous flux recovery, J_0 and $J_{fc,}$ respectively, represent the permeate flux of the virgin membrane and the membrane permeate flux after ultrasonic cleaning, $m^3\ m^{-2}\ h^{-1}$; R_m and $R_{tf,}$ respectively, represent the virgin membrane resistance and the membrane fouling resistance, m^{-1}; t is the ultrasonic cleaning duration, min.

The following equation can be obtained from the Hom-Haas model [23, 24]:

$$\ln\left(\frac{R_{tf}(t)}{R_{tf}(0)}\right) = -kC^n t^m \tag{4}$$

where k is a kinetic constant, C is the concentration of the cleaning agent, and n and m represent two different experimental constants.

After combining Eqs. (3) and (4) with considering the concentration of DI-water is constant in the ultrasonic cleaning process:

$$FR(t) = \frac{1}{1 + \frac{J_0 - J_{fc}}{J_{fc}} \exp(-kt^m)} \times 100\% \tag{5}$$

In the constant pressure filtration process, the instantaneous flux ($J(t)$) and effective filtration area of the membrane ($A(t)$) can be calculated by Eqs. (6) and (7), respectively [23, 24]:

$$J(t) = \frac{J_f\left((1-K)\exp\left(\frac{-K_b}{K_C J_0^2}\left(\left(1+2K_C J_f^2 t\right)^{1/2}\right)-1\right)+K\right)}{\left(1+2K_C J_f^2 t\right)^{1/2}} \tag{6}$$

$$\frac{A(t)}{A_f} = (1-K)\exp\left(\frac{-K_b}{K_C J_0^2}\left(\left(1+2K_C J_f^2 t\right)^{1/2}\right)-1\right)+K \tag{7}$$

where $J(t)$ and $J_{f,}$ respectively, represent the instantaneous flux in the membrane filtration process ($m^3\ m^{-2}\ h^{-1}$) and the membrane flux after the membrane filtration process ($m^3\ m^{-2}\ h^{-1}$); K represent steady membrane frontal area (m^2); K_b and $K_{C,}$ respectively, represent complete blocking resistance constant (h^{-1}) and cake resistance constant (hm^{-2}); A (t) and $A_{f,}$ respectively, represent the instantaneous available frontal membrane area in the membrane filtration process (m^2) and available frontal membrane area after the membrane filtration process (m^2), respectively.

Because ultrasonic cleaning can only remove foulants on the membrane and have no effect on foulants in the membrane pores, therefore, the cake resistance (R_c) after the ultrasonic cleaning operation can be calculated as follows:

$$R_c(t) = R_{tot}(t) - R_m - R_p(t) \tag{8}$$

where R_m is the virgin membrane resistance (m^{-1}), $R_p(t)$ is the membrane pore resistance, which consists of $R_{pB}(t)$ (the membrane pore resistance due to pore blocking (m^{-1})) and $R_{pin}(t)$ (the membrane pore resistance due to the foulants depositing and narrowing the membrane pore (m^{-1}).

According to Darcy's law and the Glen Bolton pore blocking model, the membrane pore resistance ($R_p(t)$) is inversely proportional to the available membrane frontal area (A) [23]. Therefore, the following equation can be obtained:

$$R_p(t) = \frac{a}{A(t)} + b \tag{9}$$

The ultrasonic cleaning efficiency (*CE*) can be used to calculate:

$$\mathrm{CE} = \frac{R_{tf} - R_{fC}}{R_{\mathrm{tf}}} \times 100\% \tag{8}$$

where R_{fC} is the residual cake resistance after the ultrasonic cleaning process (m^{-1}).

When the ultrasonic pressure reaches the Blake threshold of the liquid, many microbubbles will rapidly and continuously form, grow and deform, and form a high-speed microjet during the collapse of the solid boundary and remove the foulants from the membrane surface (Fig. 8).

In order to verify the practicality of the proposed model, the model predictions were fitted with the experiment data and the corresponding results (Fig. 9). The results showed that within the scope of all cleaning conditions, the predicted values of *FR(t)* by the proposed model were in good agreement with the experimental data at different ultrasonic cleaning conditions (R^2 > 0.9883). Therefore, the proposed model can be used to predict the instantaneous flux recovery (*FR(t)*) in the ultrasonic cleaning process.

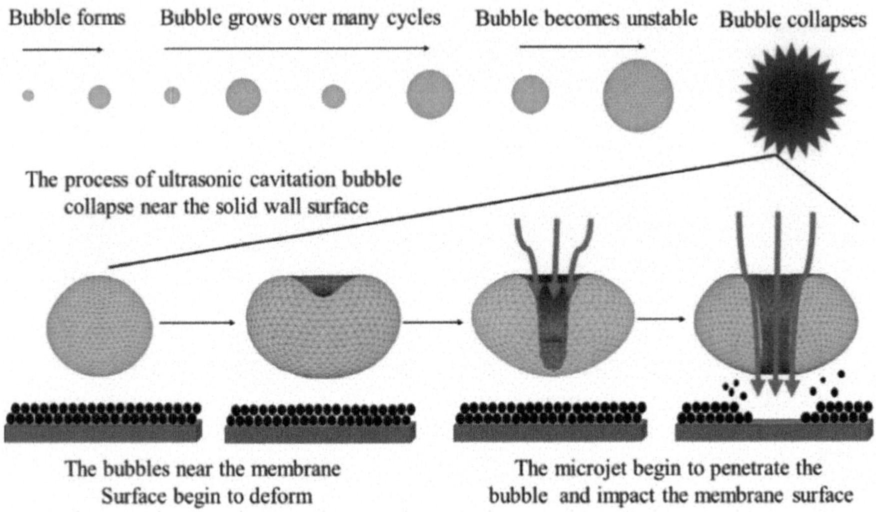

Fig. 8 The dynamic process of bubble cavitation and the membrane cleaning mechanism under the ultrasonic wave

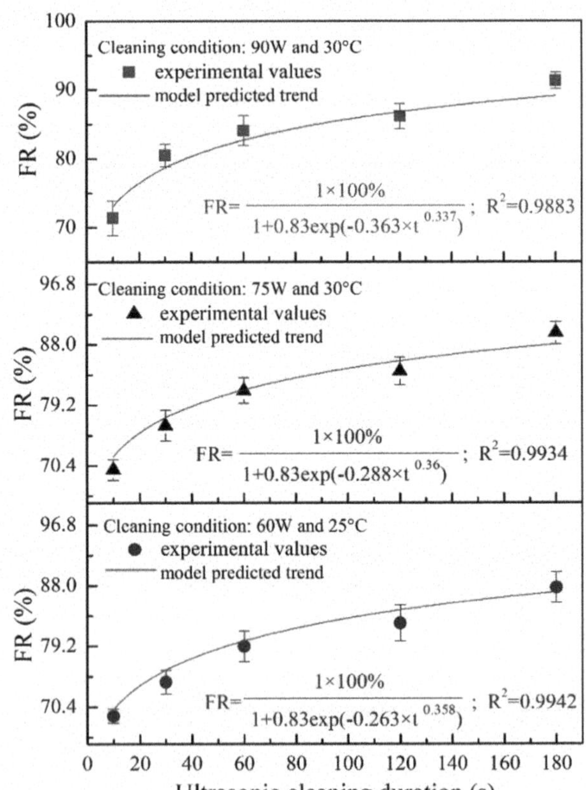

Fig. 9 Comparisons of model predictions and experiment data at different ultrasonic cleaning conditions (The membrane: 100 kDa PES UF membrane; The fouling condition: 10 mg/L HA and 1 Mm $CaCl_2$ synthetic solution at temperature = 23 ± 1 °C and TMP = 0.1 MPa and fouling 60 min; the ultrasonic cleaning conditions: 60W and 25 °C, 75W and 25 °C, 90W and 30 °C)

References

1. H.M. Kyllonen, P. Pirkonen, M. Nystrom, Membrane filtration enhanced by ultrasound: a review. Desalination **181**, 319–335 (2005)
2. M. Cai, S.J. Wang, Y.P. Zheng, H.H. Liang, Effects of ultrasound on ultrafiltration of Radix astragalus extract and cleaning of fouled membrane. Sep. Purif. Technol. **68**, 351–356 (2009)
3. X. Li, J.S. Yu, A.G.A. Nnanna, Fouling mitigation for hollow-fiber UF membrane by sonication. Desalination **281**, 23–29 (2011)
4. M. Cai, S.N. Zhao, H.H. Liang, Mechanisms for the enhancement of ultrafiltration and membrane cleaning by different ultrasonic frequencies. Desalination **263**, 133–138 (2010)
5. S. Muthukumaran, K. Yang, A. Seuren, S. Kentish, M. Ashokkumar, G.W. Stevens, F. Grieser, The use of ultrasonic cleaning for ultrafiltration membranes in the dairy industry. Sep. Purif. Technol. **39**, 99–107 (2004)
6. T. Kobayashi, X. Chai, N. Fujii, Ultrasound enhanced cross-flow membrane filtration. Sep. Purif. Technol. **17**, 31–40 (1999)
7. T. Kobayashi, T. Kobayashi, Y. Hosaka, N. Fujii, Ultrasound-enhanced membrane-cleaning processes applied water treatments: influence of sonic frequency on filtration treatments. Ultrasonics **41**, 185–190 (2003)
8. M.O. Lamminen, H.W. Walker, L.K. Weavers, Mechanisms and factors influencing the ultrasonic cleaning of particle-fouled ceramic membranes. J. Membr. Sci. **237**, 213–223 (2004)
9. W. Kim, T.H. Kim, J. Choi, H.Y. Kim, Mechanism of particle removal by megasonic waves. Appl. Phys. Lett. **94** (2009)
10. A. Maskooki, T. Kobayashi, S.A. Mortazavi, A. Maskooki, Effect of low frequencies and mixed wave of ultrasound and EDTA on flux recovery and cleaning of microfiltration membranes. Separat. Purif. Technol. **59**, 67–73 (2008)
11. E. Alventosa-deLara, S. Barredo-Damas, M.I. Alcaina-Miranda, M.I. Iborra-Clar, Study and optimization of the ultrasound-enhanced cleaning of an ultrafiltration ceramic membrane through a combined experimental-statistical approach. Ultrason. Sonochem. **21**, 1222–1234 (2014)
12. S.R. Gonzalez-Avila, F. Prabowo, A. Kumar, C.D. Ohl, Improved ultrasonic cleaning of membranes with tandem frequency excitation. J. Membr. Sci. **415**, 776–783 (2012)
13. C.C. Kan, D.A.D. Genuino, K.K.P. Rivera, M.D.G. de Luna, Ultrasonic cleaning of polytetrafluoroethylene membrane fouled by natural organic matter. J. Membr. Sci. **497**, 450–457 (2016)
14. M.O. Lamminen, H.W. Walker, L.K. Weavers, Cleaning of particle-fouled membranes during cross-flow filtration using an embedded ultrasonic transducer system. J. Membr. Sci. **283**, 225–232 (2006)
15. J.X. Li, R.D. Sanderson, E.P. Jacobs, Ultrasonic cleaning of nylon microfiltration membranes fouled by Kraft paper mill effluent. J. Membr. Sci. **205**, 247–257 (2002)
16. X.J. Chai, T. Kobayashi, N. Fujii, Ultrasound-associated cleaning of polymeric membranes for water treatment. Sep. Purif. Technol. **15**, 139–146 (1999)
17. M.J. Lujan-Facundo, J.A. Mendoza-Roca, B. Cuartas-Uribe, S. Alvarez-Blanco, Cleaning efficiency enhancement by ultrasounds for membranes used in dairy industries. Ultrason. Sonochem. **33**, 18–25 (2016)
18. S. Muthukumaran, S. Kentish, S. Lalchandani, M. Ashokkumar, R. Mawson, G.W. Stevens, F. Grieser, The optimisation of ultrasonic cleaning procedures for dairy fouled ultrafiltration membranes. Ultrason. Sonochem. **12**, 29–35 (2005)
19. D. Chen, L.K. Weavers, H.W. Walker, Ultrasonic control of ceramic membrane fouling: effect of particle characteristics. Water Res. **40**, 840–850 (2006)
20. R.S. Juang, K.H. Lin, Flux recovery in the ultrafiltration of suspended solutions with ultrasound. J. Membr. Sci. **243**, 115–124 (2004)
21. D. Chen, L.K. Weavers, H.W. Walker, Ultrasonic control of ceramic membrane fouling by particles: effect of ultrasonic factors. Ultrason. Sonochem. **13**, 379–387 (2006)

22. H.J. Luo, Z. Wang, A new ultrasonic cleaning model for predicting the flux recovery of the UF membrane fouled with humic acid. J. Environ. Chem. Eng. **10** (2022)
23. G. Bolton, D. LaCasse, R. Kuriyel, Combined models of membrane fouling: development and application to microfiltration and ultrafiltration of biological fluids. J. Membr. Sci. **277**, 75–84 (2006)
24. L. Hou, Z. Wang, P. Song, A precise combined complete blocking and cake filtration model for describing the flux variation in membrane filtration process with BSA solution. J. Membr. Sci. **542**, 186–194 (2017)

The manufacturer's authorised representative in the EU is Springer Nature Customer Service Centre GmbH, Europaplatz 3, 69115 Heidelberg, Germany. If you have any concerns regarding our products, please contact ProductSafety@springernature.com

Printed and bound by CPI Group (UK) Ltd, Croydon, CR0 4YY
30/09/2026
02225304-0001